Living in a Garden

THE GREENING OF SINGAPORE

50 years
of
Greening Singapore
Our City in a Garden

Living in a Garden

THE GREENING OF SINGAPORE

Text by Timothy Auger

NATIONAL PARKS
LET'S MAKE SINGAPORE OUR GARDEN

edm EDITIONS DIDIER MILLET

Front Cover: A block of flats in Ang Mo Kio as viewed through the canopy of one of Singapore's many roadside trees.

Pages 2–3: Children playing in one of Singapore's many parks.

Pages 4–5: Little Tern (*Sterna albifrons*) and its chick. These birds can be seen at Sungei Buloh Wetland Reserve and Pulau Ubin.

Pages 6–7: A Kapok Tree (*Ceiba pentandra*) (right) and a Mengkulang (*Heritiera elata*) (left), two of the 34 Heritage Trees found in the Singapore Botanic Gardens.

Page 8: Even in the heart of the city, there are many parks where people can exercise or take a break from the daily grind and relax amid nature.

Pages 10–11: Hornbills are an increasingly common sight in Singapore. At the start of the breeding season, a male Oriental Pied Hornbill (*Anthracoceros albirostris convexus*) checks out a tree cavity before inviting the female to nest and breed.

Pages 12–13: Two Collared Kingfishers (*Todiramphus chloris*), a bird species commonly seen in the city.

Published in 2013 for National Parks Board by
Editions Didier Millet
121 Telok Ayer Street #03-01
Singapore 068590
Tel: (65) 6324 9260
Fax: (65) 6324 9261
Email: edm@edmbooks.com.sg
www.edmbooks.com

EDM Editorial Team
Editorial Director: Douglas Amrine
Editors: Rachael Morris and James Lui
Designer: Tan Seok Lui
Production Manager: Sin Kam Cheong
Photo Research: Amandas Ong

Colour separation: PICA Digital Pte Ltd
Printed in Singapore by Tien Wah Press Pte Ltd

ISBN 978-981-4385-24-4

CONTENTS

FOREWORD
by
PRESIDENT
15

1
THE BIRTH OF
A GARDEN CITY
18

2
BREAKING
NEW GROUND,
REDISCOVERING
ROOTS
56

3
WHO ARE
THE PARKS
FOR?
86

4
·
ANYONE CAN
BE A GARDENER
·
114

5
·
WORLDS
OF NATURE
·
124

6
·
BIODIVERSITY,
A SHARED
HERITAGE
·
156

7
·
CITY IN
A GARDEN
·
168

APPENDICES
·
198

FOREWORD

Singapore is recognised today for being a clean and green city. Our lush landscape is an important facet of our identity. The garden environment and green spaces contribute to an enjoyable and liveable environment in which Singaporeans live, work and play. Visitors to Singapore are struck by the sight of tree-lined highways, manicured lawns and smartly-pruned hedges that greets them. Our reputation as a City in a Garden enhances Singapore's attractiveness as a destination for tourists, foreign businesses and global talents.

The many vibrant parks and gardens, the biodiversity of our nature reserves, and the extensive network of park connectors and green corridors that we now enjoy are our natural heritage. These are the results of deliberate and sustained efforts that began 50 years ago. The resolve, fortitude and industry of the many stakeholders and partners in creating and maintaining our country's greenscape are encapsulated in this publication. I hope that the book will inspire many more Singaporeans to contribute to this national endeavor. It is the responsibility of all Singaporeans to maintain and conserve Singapore's natural heritage for future generations to enjoy for many more years to come.

Dr Tony Tan Keng Yam
President of the Republic of Singapore

Cratoxylon formosum
(Mempat)
Planted by the
Prime Minister
Mr. Lee Kuan Yew
16.6.63

Let's make the city green!

The Straits Times announced, "Lee begins the tree campaign." On 16 June 1963, Singapore's then prime minister, Lee Kuan Yew, toured the Ulu Pandan constituency. At that time, the junction of Farrer Road with Holland Road was a roundabout, known as Farrer Circus. There, he planted a tree.

As reported in the newspaper, the ceremony was the start of an island-wide tree-planting campaign. The purpose was to bring rain as the nation was suffering from drought, which was to last 10 months. Water from pipes was rationed and even cut off for up to 12 hours at a time. Health warnings went out regarding food stalls that did not use boiled water for food preparation.

Later, the tree-planting campaign became part of a plan for a complete physical, social, cultural and economic makeover. Singapore would be a "garden city beautiful with flowers and trees, and as tidy and litterless as can be," the prime minister said in May 1967.

The challenge was huge. In the 1960s, flies hovered over hawker stalls and piles of rubbish. Only 60 percent of the city's refuse was regularly cleared. People kept animals in their backyards. Cows grazed on roundabouts. Drains were clogged. Duck and pig farms on the riverbanks, as well as bumboats dripping oil, contributed to pollution of the waterways, which stank. At the same time, the pace of urban development was in danger of turning the city into a concrete jungle.

Who could have imagined in the 1960s what Singapore would look like 50 years on?

Today, visitors to Singapore admire the greenery. The streets and waterways are clean. The network of parks, gardens and nature reserves, and the thriving native biodiversity, would have been dismissed as a fantasy only a few decades ago.

"Without the greening effort, Singapore would have been a barren, ugly city. There would have been a few trees, planted haphazardly here and there, but there would have been none of the planning or the care and maintenance that sustain our greenery today."
Lee Kuan Yew (2012)

● Prime Minister Lee Kuan Yew plants a Mempat tree (*Cratoxylum formosum*) to kick off a tree-planting campaign in 1963.

1

THE BIRTH OF A GARDEN CITY

The population of newly independent Singapore faced challenges. Homes and jobs were scarce. Yet, the government made greening a priority. The first tree-planting campaign was just the first step in a far-reaching transformation. It would distinguish Singapore from other fast-developing cities and encourage international businesses to invest in the local economy. Most importantly, it would make Singapore a pleasant place in which to live and raise a family.

● Toa Payoh Town Garden is the setting for a wedding photograph in the 1970s.

Where did the forests go?

The disappearance of much of the primary vegetation in Singapore was the result of clearance for plantations, rather than modern development such as the building of new towns at Jurong, Toa Payoh, Woodlands or Tampines.

The chief villain was gambier (*Uncaria gambir*), a plantation crop that Chinese settlers had grown on the nearby island of Bintan from the mid-18th century. There were already several gambier plantations on Singapore itself when Sir Stamford Raffles arrived in 1819. Thirty years later, gambier covered the greater part of the entire cultivated area.

Pepper was often grown with gambier. Nutmeg was also widely grown until a blight affected the crop in the mid-19th century.

The population of Singapore was growing. Then, as it is now, securing water supply was an urgent priority. Reservoirs replaced some of the forests, beginning with the MacRitchie Reservoir.

After the decline of the gambier industry towards the end of the 19th century, rubber replaced gambier in many areas. Pineapple and coconut were grown extensively. Some land was used for other forms of agriculture, and some became overgrown with lallang.

During the first half of the 20th century, change was gradual. Although the built-up area was growing, as recently as the 1950s and 1960s some 20,000 farms still occupied about a quarter of Singapore's land area. The farmers grew vegetables, fruits such as coconuts, and rubber. There were still a few small plantations growing pineapples, tobacco and spices. Rearing of pigs and poultry was combined with vegetable farming. Orchid growing was a substantial industry for a time.

From the 1960s, the transformation was dramatic. Farmers had to make way for infrastructure projects, housing and industry. Housing development, started under the Singapore Improvement Trust in the 1930s, speeded up after the creation of the Housing and Development Board (HDB) in 1960. Families moved out of rural kampongs into housing estates. Western Singapore was transformed by the Jurong Industrial Estate. Reclamation redrew Singapore's coastline, increasing the land area.

Today, despite the changes, some 5 percent of Singapore's primary vegetation remains. There are also larger areas of mostly regenerating secondary rainforest, stretches of mangroves and unaltered parts of the shoreline, which host diverse marine life.

● Above: The multi-tiered forest canopy at the Bukit Timah Nature Reserve, which is protected today.

● Opposite: Much of the primary forest was cut down in the 19th century to make way for plantation crops.

"Why bother about a few trees? It is true that a few trees, when cut down, have a comparatively small value; but it is not their individual value as dead timber with which we should be concerned."

R.E. Holttum, Singapore Botanic Gardens director, 1926–1949 writing in *The Straits Times* (1950)

In the Jungle

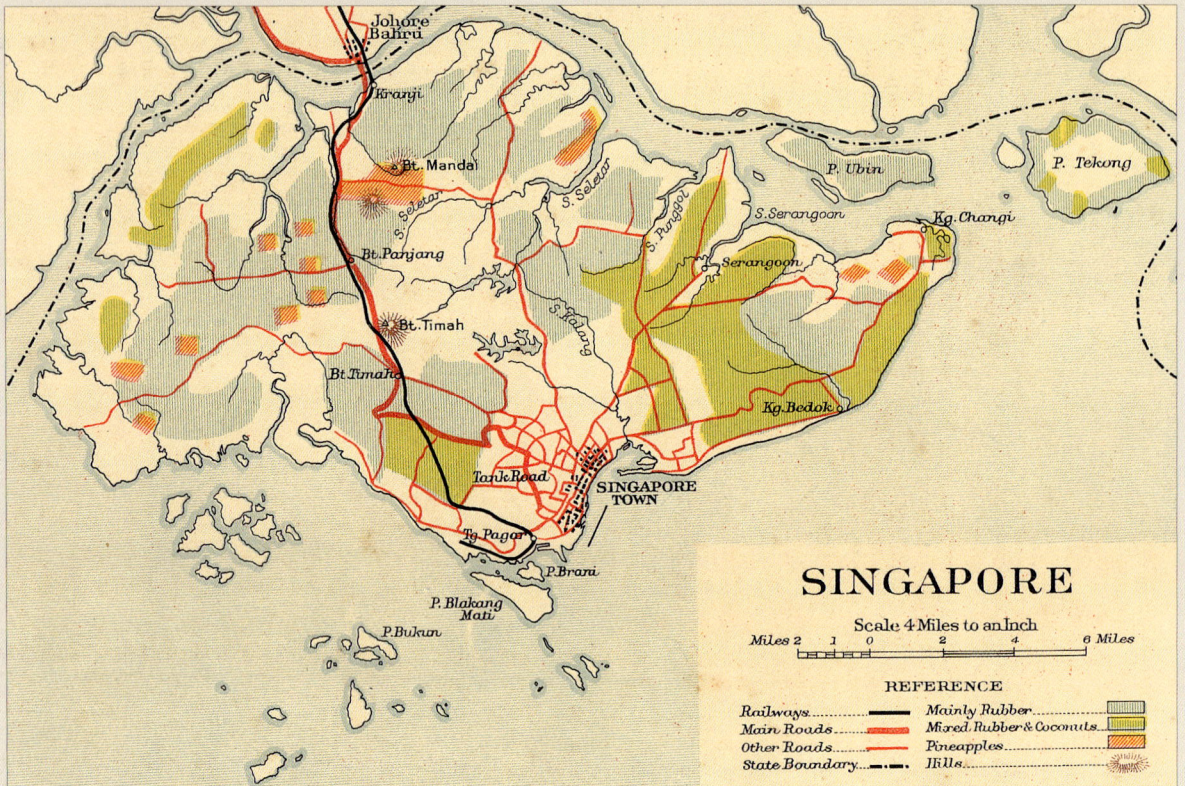

SINGAPORE

Scale 4 Miles to an Inch

Miles 2 1 0 2 4 6 Miles

REFERENCE

Railways ....   Mainly Rubber ....
Main Roads ....   Mixed Rubber & Coconuts ...
Other Roads ....   Pineapples ....
State Boundary ....   Hills ....

● Map showing the land covered by the main plantation crops in the 1930s. Much of the primary forest had already disappeared by this time.

Gambier

Gambier (*Uncaria gambir*) is a climbing plant native to the Malay Peninsula. Plantation workers boiled its leaves to extract tannin, which was exported to China and Java for tanning and dyeing. After clearing forest to grow gambier, they had to clear the same amount of land again for the firewood needed to boil the leaves. After about 15 years, the soil contained few nutrients, and there would be no more timber immediately at hand, so the planters moved on to another piece of virgin land. The impact was catastrophic for both the forest and the animals that lived there.

Rubber

Rubber (*Hevea brasiliensis*) brought prosperity to Malaya, including Singapore, in the first half of the 20th century. It became a plantation crop thanks to the efforts of Henry N. Ridley, the first scientific director of the Singapore Botanic Gardens. In the photograph above, he is demonstrating the latex tapping process. By 1920, Malaya was meeting half the world's rubber needs, with Singapore as the main processing and trading centre. From the 1950s onwards, because of sagging prices and land scarcity, the Singapore rubber industry declined.

Nutmeg

The seed of the nutmeg tree (*Myristica fragrans*) was very precious 200 years ago – it was the source of nutmeg and mace, spices used in cooking. Following Sir Stamford Raffles' arrival in Singapore, the tree was brought from the Moluccas, also known as the Spice Islands. Thirty years on, Singapore's population of nutmeg trees had grown to some 7,000. The industry declined after a nutmeg-beetle blight ravaged the plantations in 1850. Nutmeg is one of the crops that gave Orchard Road its name.

"View of the jungle, Singapore": the forest as it must have appeared in the mid-19th century.

The forest that Raffles found

Primary vegetation is what grows on land that man has never cleared. Secondary vegetation is what grows after land clearance.

Two centuries ago, lowland tropical rainforest covered the island (there are still remnants in the Central Catchment and Bukit Timah Nature Reserves, and the Singapore Botanic Gardens), together with some areas of freshwater swamp forest (some has survived at Nee Soon).

Most of the primary rainforest trees are members of the Dipterocarp family. They are slow-growing and majestic. However, there were representatives of many other families of trees too. The forest canopy can reach a height of 50 metres. The greatest recorded height for a Dipterocarp tree is 88 metres, taller than a 20-storey building.

Along stretches of seashore and the river estuaries grew mangroves, of which only remnants now remain.

Pioneers in conservation

Nathaniel Cantley, the then superintendent of the Singapore Botanic Gardens, surveyed the forests in 1882. He set out conservation guidelines that still make good sense and outlined forest reserves, including Bukit Timah and some mangrove areas, on the map.

In 1935, the forest reserves, dismissed as economically unproductive, were officially abolished. To the rescue came R.E. (Eric) Holttum (right) and E.J.H. Corner (far right), director and assistant director at the Singapore Botanic Gardens, who persuaded the authorities to reinstate the reserves at Bukit Timah, Kranji and Ulu Pandan in 1939, "to provide areas for research, education, recreation and as samples of the country's biographic history and heritage".

Bukit Timah was a rich source of granite. By 1948, one quarry had reached the edge of the Bukit Timah reserve and others were alarmingly close.

The preservation of forest at Bukit Timah and in the adjoining catchment area was under threat. A new law came into force in 1951, protecting the existing reserves plus the catchment area and Labrador Cliff, as it was then called.

The importance of greening

By the 1970s, urban development was going at full tilt. Nevertheless, Lee Kuan Yew made it clear that greening was a priority.

Why was it so important to plant trees? Lee's idea was to make Singapore distinct from its neighbours. As he wrote in his memoirs, "One arm of my strategy was to make Singapore into an oasis in Southeast Asia, for if we had First World standards, then businessmen and tourists would make us a base for their business and tours of the region." This emphasis on long-term attention to appearances was key to the transformation of Singapore into a garden city then, and it is still important.

The government gave the task of greening Singapore to the Parks and Recreation Division, formed in 1974 within the Public Works Department under the Ministry of National Development (MND). At its helm was Wong Yew Kwan, who transferred from the Primary Production Department (PPD). In 1976, when the Parks and Recreation "Division" became a "Department" (PRD), Wong became the first Commissioner of Parks and Recreation. His team was small, some 40 strong. It included only a handful of local graduates and one graduate botanist, although a number of recruits had Taiwanese degrees in horticulture or agriculture.

● Above: Shenton Way during redevelopment in 1974. Large expanses of bare concrete were not yet softened by vegetation.

● Opposite: Toa Payoh Town Garden (as it was originally known) during the construction phase in 1973 (top) and as it is today (below), an oasis of green.

"When I went to Britain to study I saw the Royal Botanic Gardens at Kew. I thought, 'If they can bring a tropical forest to a temperate country, it would be stupid for us not to manage our own natural resources to bring greenery into our city. We've got the natural resources; we just have to pay a certain amount of attention to details.'"

Lee Kuan Yew (2012)

Everybody plays a part

The greening of Singapore required public cooperation. A month-long "Keep Singapore Clean" campaign began in October 1968, to counter littering. The message was spread through the schools and the mass media. Other campaigns followed in 1970: "Keep Singapore Clean and Mosquito-Free", "Keep Singapore Clean and Pollution-Free" and "Keep Singapore Pollution-Free".

The first official Tree Planting Day was Sunday, 7 November 1971. Earlier that week, schoolchildren had planted 600 trees. On the day itself, volunteers planted 8,400 trees and 21,677 shrubs and creepers. Not all of them flourished according to plan, as Wong Yew Kwan recalls. "Dr Goh Keng Swee (as acting prime minister) planted a Rain Tree on Mount Faber. When the park was redeveloped some years later, the root zone was affected, and the tree had to be containerised to preserve it."

Even the trees planted by Lee Kuan Yew occasionally underperformed, as he himself recalled in 2012. "After one Tree Planting Day, I passed by and found that the tree I had planted had died. I realised that just planting isn't good enough. You have to nurture the tree and make it grow sturdy and strong. That's what spurred me on."

On 4 November 1990, Tree Planting Day became part of Clean and Green Week, forerunner of today's "Clean and Green Singapore" campaign. Singaporeans today can still join in tree-planting activities through the Plant-A-Tree programme.

● Below: St Andrews School students take part in a tree-planting exercise in 1971.

● Left: Dr Goh Keng Swee planting a tree in the Kreta Ayer constituency in the 1970s.

● Above: Aljunied residents help political secretary to the Ministry of Education, Mohamed Ghazali, plant a Yellow Flame tree in Jalan Balam, 1970.

Fruit trees everywhere – for a time

The 1970s and 1980s saw campaigns for the planting of fruit trees in public housing estates, condominiums, government institutions, parks and open spaces. In one year, 1984, over 3,500 fruit trees were planted on HDB estates. By 1996, the cumulative total was 44,000.

The idea was to strengthen civic consciousness and social discipline. There was also an educational purpose: to teach children, no longer brought up in kampongs, to recognise different kinds of fruit.

Dr Goh Keng Swee proposed the growing of fruit trees in military camps. According to Dr Chua Sian Eng, appointed parks commissioner in 1983, Dr Goh had seen the GD ("General Duties") recruits, just out of school, hanging around the garages doing nothing. "Let's make them do some work," Dr Goh said. So work they did (below left).

Soon, some fruit showed marks where children had tested them for ripeness. Some were stolen. The prime minister was philosophical, Dr Chua recalls, advising that the number of fruit trees be increased until the thieves just got bored with stealing.

Long-term results were mixed. Contractors did not find harvesting the fruit commercially attractive. Fruit would drop, rot and attract flies. Many fruit trees have since disappeared because of redevelopment and pressure on land, although some housing committees and community centres continue the tradition. Trees bearing popular fruit like durian and jackfruit (*Artocarpus heterophyllus*) (below right) can still be seen around neighourhoods.

Moving as one

For the greening of Singapore to succeed, the whole of government had to pull in the same direction. Planting had to be a priority and not sidelined. Ensuring this was, and still is, the job of the Garden City Action Committee (GCAC), originally formed in the late 1960s. Its members represent all the ministries and statutory boards contributing to the greening effort. It hears reports on operational and maintenance issues, streetscape greenery, developments affecting the parks, the Singapore Botanic Gardens and the nature reserves – no aspect of the greening of Singapore escapes its notice.

In 1978, according to the first parks commissioner Wong Yew Kwan, the prime minister felt that the process of turning Singapore into a garden city needed a boost. He called a meeting and made it clear to civil servants that they should channel increased resources to it. By 1980, the budget was nearly 10 times what it had been in 1973, after inflation is taken into account. More staff joined the team, including several who had returned to Singapore having earned degrees in the United Kingdom, Australia, New Zealand or Malaysia.

The GCAC's role has evolved. In the early days, the drive behind Singapore's transformation came from the top, in the form of instructions communicated through the Committee. Today, the notion of a garden city is firmly embedded in the thinking of the PRD's successor, the National Parks Board (NParks) and the public and private organisations whose work affects the environment. There is a critical mass of expertise and commitment that did not exist previously. Even more importantly, the public sees greenery and conservation as priorities. While the GCAC still has an active role, the creation of a "city in a garden" has acquired its own momentum.

A view over forest trees: Telok Blangah Estate and the central business district as seen from Mount Faber.

What to plant?

Up to the end of the 1970s, PRD officers chose trees mostly for shade, rather than for colourful foliage or flowers. Some of the trees were "native" – they grew in the primary forests. Others were introduced from overseas, although the range available then was smaller than that of today. Seeds collected abroad were put into quarantine; cuttings were treated to eliminate pests and diseases. After that, the newly introduced species that were successful could be grown in bulk.

Not every native tree grows well in the city. In natural conditions, trees spend much of their lives in the forest understorey. When, after growing to their full height, mature trees reach the end of their life and need to be replaced, younger trees grow up and take their place in the canopy.

In the city, however, trees are exposed to open skies from the beginning, and this affects the way some of them grow.

Trees need water. Although Singapore gets some 2.5 metres of rain every year, in built-up areas much of it flows straight into the drains. So some trees from tropical countries with climates drier than Singapore's proved to be more suitable for urban conditions.

Sea Gutta (*Planchonella obovata*) (or Nyatoh Laut in Malay) is native to Southeast Asia, Australia and the Pacific Islands, where it grows near sandy shores and rocky headlands. It was once a common tree in Singapore but almost disappeared when the coastlines were modified by reclamation. It is a hardy tree, tolerant of salt-spray, seawater inundation and urban pollution. The wood can be used for carvings, furniture, house posts and saltwater pilings. It is a member of the same family (Sapotaceae) as Chiku (*Manilkara zapota*) and Gutta Percha (*Palaquium gutta*). Like these trees, it exudes a milky sap when wounded. Today, it is commonly planted along roadsides – most obviously North Bridge Road and New Bridge Road.

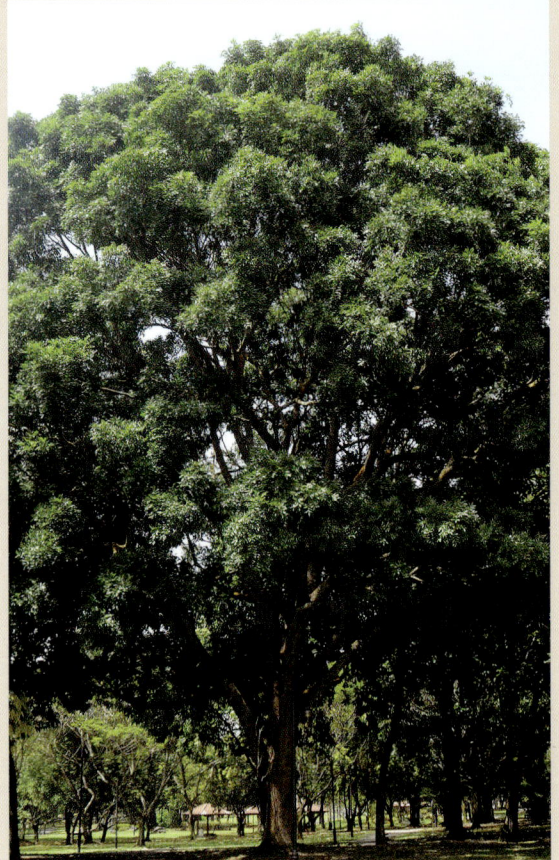

The **Senegal Mahogany (*Khaya senegalensis*)** is a native of tropical West Africa. Capable of reaching 25 metres in height, it has surprised NParks officers with its speed of growth – within 20–30 years it has become one of the biggest trees in Singapore's parks and along roadsides. The Senegal Mahogany is closely related to the true mahoganies (genus *Swietenia*). Its wood is used in furniture, cabinetwork and other woodwork.

The **Saga** (*Adenanthera pavonina*) is not strictly native, having originated in other parts of tropical Asia. However, it was introduced to Singapore a long time ago and has since become naturalised. Growers of coffee, cloves and rubber in past times used it as a fast-growing shade tree for their plantations. It is less commonly used along roads now because it sprouts profusely, adding weight to the tips of branches and making them more susceptible to damage in high winds. In times past, traders of gold and silver used the decorative, bright-red seeds often found beneath the tree as weights – in fact, the very name "Saga" has been traced to the Arabic word for "goldsmith". In Malaysia, Proton even named a car after it!

The **Angsana** (*Pterocarpus indicus*) was a favourite shade tree during Singapore's first greening phase in the 1960s and 1970s. It has a dense, dome-shaped crown with attractively drooping branches. A native of India and Southeast Asia, it is easily propagated and transplanted. It grows quickly and can reach 30–40 metres in height, developing buttressed roots. Sadly, much of the island's Angsana population was destroyed by Angsana wilt (see below).

The fast-growing **Sea Almond** or **Ketapang** (*Terminalia catappa*), which occurs naturally along the seashores of tropical Asia and Australasia, is a salt-tolerant tree, up to 25 metres in height, with long spreading branches. Its branches grow in wide spreading tiers, creating a pagoda-shaped crown. Twice a year, before they fall, the huge leaves turn red and yellow, creating an autumnal effect. It develops buttressed roots that kids love to play among during hide-and-seek. The fruit floats on water, its dispersal aided by sea currents.

Angsana wilt
A century ago, Angsana trees in Singapore were found to be suffering from a fungal disease. The spores of the fungus are carried by ambrosia beetles. By 1991, Angsana wilt disease was killing some 28 trees a month. One theory is that the damage caused by lightning strikes makes some trees more vulnerable to invasion by the beetle. Fungicide treatment (above) was tried in the past, with limited success. Some Angsana strains are proving more resistant to the disease.

Cengal Pasir (*Hopea odorata*) is one of the few Dipterocarp species planted beside roads. It grows straight and tall, up to 25–30 metres. It is native to Southeast Asia. Its natural habitat is near rivers and streams in lowland forests. In the rainforest, Dipterocarps exhibit a behaviour called "mast fruiting", where all the trees of one or more Dipterocarp species produce fruit at once, showering the forest floor with little helicopter-shaped seeds. It remains to be seen if the Dipterocarps planted along our roads will behave similarly. The timber is suitable for making boats and canoes.

The **Yellow Flame** (*Peltophorum pterocarpum*) occurs naturally in Southeast Asia, China and tropical Australia. It grows fast, reaching 15–20 metres in height. During the flowering season, the crown is covered with brilliant yellow blossoms. Although its natural habitat is mangrove forest and coastlines, it adapts well to life by the roadside and in parks and gardens. Like the Angsana, the Yellow Flame has been a feature of Singapore's built-up areas for many years.

One of the most familiar trees in Singapore today, the **Rain Tree** (*Samanea saman*) in fact hails from tropical America – colonial authorities took it to other parts of the world, including Singapore, back in the 19th century. Its characteristic shape makes it suitable for planting along roads, including expressways, where it can create a "green tunnel". Many Rain Trees are hosts to epiphytes such as Pigeon Orchids, Stag's Horn Ferns and Bird's Nest Ferns. The name comes from the leaves' habit of folding up at night or during overcast, rainy days.

The **Sea Apple** (*Syzygium grande*) (Jambu Laut in Malay) is native to Singapore, Malaysia and elsewhere in Southeast Asia. It is salt tolerant and grows naturally in coastal settings. A favourite wayside tree in Singapore since the 1900s, it can grow up to 25–30 metres. It is said to be relatively fire-resistant – it served as a fire-break in days when lallang grass growing on former plantation land sometimes caught fire. It is a member of the same genus as the popular kampong fruits Jambu (*Syzygium aqueum*) and Jambu Bol, or Rose Apple (*Syzygium jambos*).

The **Tembusu** (*Fagraea fragrans*) is a native of Singapore and the surrounding region. It can live to a great age and reach 35 metres in height – one tree, in the Singapore Botanic Gardens, may be more than 200 years old, and a small number of others, in the Istana garden, are estimated to be older than the Istana itself. It is slow growing and does not mind the poorly aerated clay soil covering much of Singapore. The hard, heavy timber is no longer used commercially, but in the past made good chopping boards. The deeply fissured bark is distinctive. Sometimes branches can grow out horizontally, as illustrated on the Singapore five-dollar note issued in 1999. More commonly, they grow horizontally then make a 90-degree turn towards the sky once they reach the edge of the canopy. New shoots emerge from the branches at equal intervals, always in opposite pairs, and each pair is oriented exactly at a right angle to the previous pair. Unusually for trees in Singapore, it has distinct flowering seasons, two a year. Moths, and sometimes bats, pollinate the flowers at night; bats and birds (especially starlings) eat the fruit and disperse the seeds.

How to move a tree

Trees take time to grow and Singapore needed quick results. The solution was "instant trees". Instead of using small container-grown saplings, NParks officers transplanted young trees, allowing large areas to be greened up quickly. Rain Trees and Angsanas were particularly successful.

The demand for instant trees was a commercial opportunity for the nursery industry, which started to supply saplings in large numbers, many of them grown in Malaysia. After the nurserymen identified some tree species that could survive in an urban environment, NParks officers would send them lists of those needed for upcoming park and streetscape developments. The nurserymen would then pre-grow trees to a semi-mature state in anticipation. Today, although newly introduced species are tested and grown locally to begin with, some 70 percent of the trees planted in Singapore have spent their early years outside the country.

When maintenance staff move a tree, they prune it back if necessary and lift it out of the ground complete with its root ball (below). It is possible to transplant a mature tree, but this is cumbersome and expensive. On the other hand, saplings transplanted into a city environment when they are young and small do not create a lush green effect immediately. So saplings are sometimes planted out first in tree banks on vacant land and moved to their final positions when they have grown bigger.

Researchers in Singapore have taken the "instant tree" concept further, growing trees in extra-large containers, 1.5 metres high and 2 metres in diameter. This approach makes sense, say, when road widening is planned for the future. Trees in containers can go into position temporarily and move somewhere else later. The containers are perforated – fine mesh over the holes lets moisture, nutrients or even root hairs through, but not structural roots. Early tests have shown that trees can grow in these large "pots" to a trunk diameter of over 1 metre. Trees that tolerate water around their roots, such as Sea Gutta, have done best. Because of the need for stability, this approach would not suit species with large crowns, such as Angsanas or Rain Trees, but with the right choice of tree it can be very useful when space is limited.

A more colourful city

In 1979, the PRD began a new phase in the development of Singapore as a garden city. There was a need for more variety. This came in the form of colourful flowers and foliage.

The shift of emphasis was dramatic. By the end of 1980, the PRD had planted 56,000 flowering trees. In 1984, nearly two-thirds of all the new plantings in Singapore were flowering shrubs. A global search was launched to find the right sort of plants. The parks commissioners and their colleagues contacted their counterparts, some of whom were old friends and classmates, in botanic gardens and other scientific institutions in places as far afield as Puerto Rico, Hawaii and Canada. They travelled to Brazil, South Africa and Mauritius. Closer to home, knowledge was shared with botanic gardens in Sri Lanka and Malaysia, and in Indonesia, where the gardens at Bogor, south of Jakarta, date back to the 18th century. The result was a wealth of advice and new plants to try.

Careful study of climate differences narrowed down the choice. The Singapore Botanic Gardens had already set up a plant introduction unit to test cuttings and seedlings. When enough plants of a new species were established, the unit would contact the maintenance staff, who would try out the plants on vacant land, on roadside verges or in some of the parks. Planted on road dividers, traffic islands and pedestrian bridges, many of these new arrivals flourished in the sun.

The emphasis on colour and variety has been maintained by NParks to this day, while every effort is made to plant more native species.

Central and South America

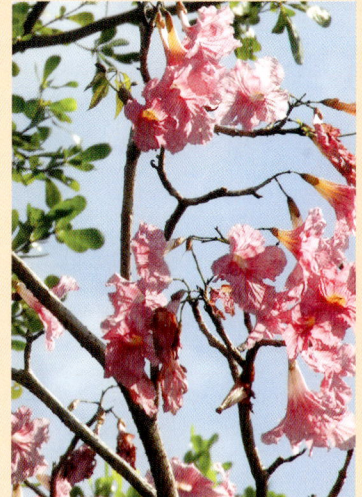

● **Bougainvillea (*Bougainvillea sp.*),** a woody vine native to Brazil and neighbouring countries, grows vigorously in our environment. Many unsuspecting gardeners have planted it beneath a tree, only to find the tree soon covered in Bougainvillea. The brightly coloured parts of the plants, so attractive on roads and bridges, are actually modified leaves, called "bracts"; the flower itself measures just 5 mm across.

● **Frangipani (*Plumeria sp.*)** is a native of tropical America. It grows up to 5–8 metres in height, with a short trunk, many branches and grey, knobbly bark. It blooms all year round in Singapore. The flowers are 5–6 cm in diameter in shades of red, pink, orange and white. They are showy and fragrant, to attract sphinx moths, but, interestingly, do not contain any nectar. So, moths are "tricked" into pollinating the plant but do not get any reward for their efforts.

● The **Trumpet Tree (*Tabebuia rosea*)** may grow to more than 30 metres in height. After a prolonged dry spell, usually in April and August each year, Trumpet Trees across Singapore will flower all at once, carpeting the streets with pink petals a few weeks later. The large, showy flowers are trumpet-shaped, hence the name. The Trumpet Tree is native to an area of Central and South America stretching from Mexico to Venezuela and Ecuador.

Where did the colour come from?

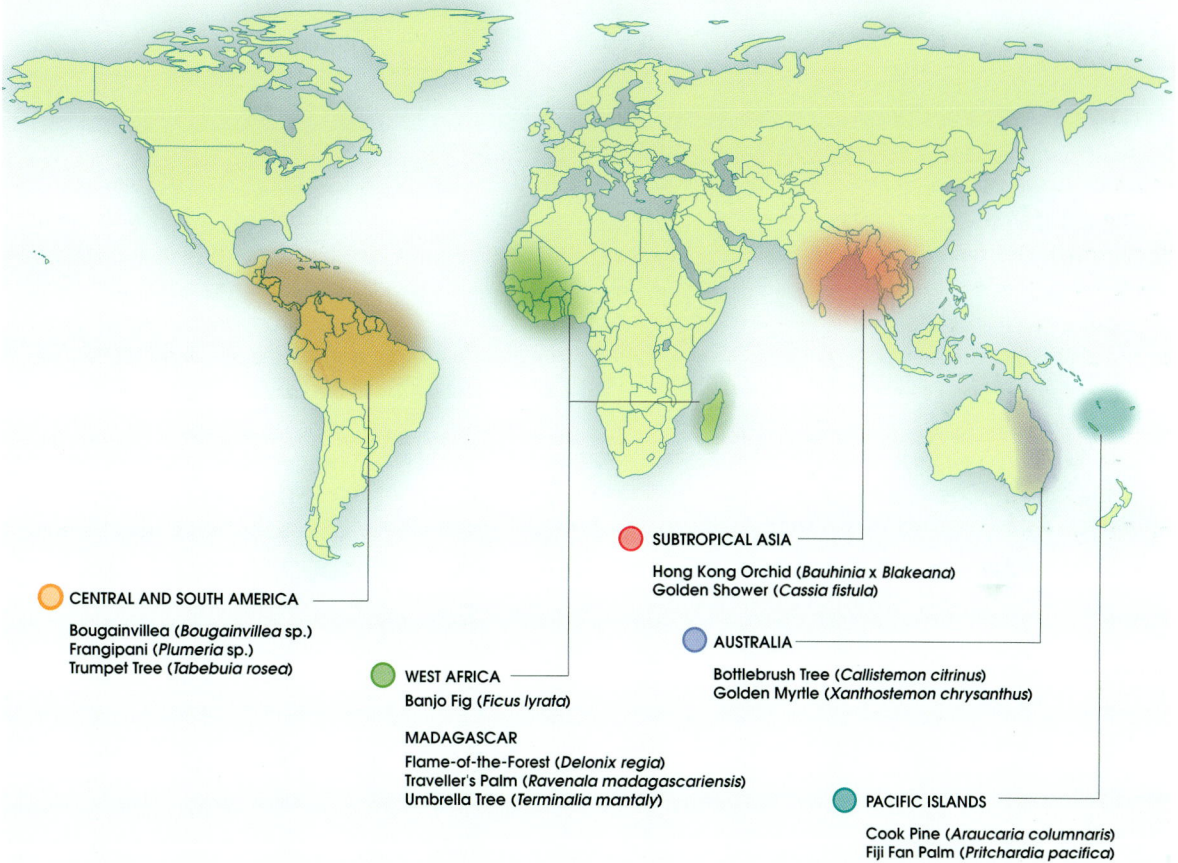

CENTRAL AND SOUTH AMERICA

Bougainvillea (*Bougainvillea* sp.)
Frangipani (*Plumeria* sp.)
Trumpet Tree (*Tabebuia rosea*)

WEST AFRICA

Banjo Fig (*Ficus lyrata*)

MADAGASCAR

Flame-of-the-Forest (*Delonix regia*)
Traveller's Palm (*Ravenala madagascariensis*)
Umbrella Tree (*Terminalia mantaly*)

SUBTROPICAL ASIA

Hong Kong Orchid (*Bauhinia* x *Blakeana*)
Golden Shower (*Cassia fistula*)

AUSTRALIA

Bottlebrush Tree (*Callistemon citrinus*)
Golden Myrtle (*Xanthostemon chrysanthus*)

PACIFIC ISLANDS

Cook Pine (*Araucaria columnaris*)
Fiji Fan Palm (*Pritchardia pacifica*)

West Africa and Madagascar

● The **Umbrella Tree** (***Terminalia mantaly***) (right) is a native of Madagascar. It can be fast-growing, reaching 10–20 metres in height. The crown grows in symmetrically arranged tiers. The bark is pale grey and, smooth, with raised brownish streaks or spots. The tree produces buttress roots as it matures. It rarely produces fruits or flowers in Singapore. Its tiered form makes it a favourite of landscape architects, providing shade while allowing building facades to peek through. The 'Tricolor' cultivar is especially popular for its mix of red, green and white leaves.

● The **Banjo Fig** (***Ficus lyrata***) (left), from tropical West Africa, gets its name from the fiddle-like shape of its leaves. In the open it can reach 15 metres in height. The fruits are speckled with white dots. Figs have an interesting symbiotic relationship with wasps. Each fig species is pollinated by a specific species of wasp. However, since the Banjo Fig's pollinator is only found in Africa, the Banjo Figs in Singapore are believed to produce figs without pollination.

● The **Golden Shower** (*Cassia fistula*) comes from India, Sri Lanka and tropical Asia. It is hardy and relatively easy to grow and reaches up to 18 metres in height. The flowers are large (3–5 cm across), brilliant golden-yellow in colour and grow in hanging clusters. They attract butterflies and other insects. This is the national tree of Thailand (where the yellow colour symbolises royalty) and the state flower of Kerala. It is said to have medicinal properties for treating a range of medical conditions, from pimples, burns, wounds and colds to cancer.

● The **Hong Kong Orchid** (*Bauhinia* x *Blakeana*) is a small hybrid tree, originating in Hong Kong, which grows up to 10 metres in height. It is very suitable for small gardens as well as parks and open spaces. It grows well in full sun. The leaves are shaped like butterfly wings (hence another common name, "Butterfly Tree"). The deep-pink flowers, resembling orchids, are large and showy and attractive to sunbirds. The flower appears on the official flag of Hong Kong. It is, strictly speaking, not an orchid but a legume (a member of the bean family).

Madagascar

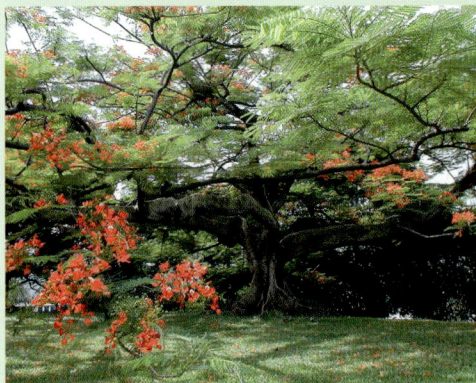

● The **Traveller's Palm** (*Ravenala madagascariensis*) originated in Madagascar. The light-green leaves grow in two ranks to form an instantly recognisable, fan-shaped crown, reaching 12–15 metres in height. It is, strictly speaking, not a palm, being closely related to the Bird of Paradise plant (*Strelitzia reginae*). In times past, rain collected in the sheaths at the leaf bases provided an emergency water supply for thirsty travellers, hence the common name.

● The crown of the **Red Flame Tree**, or **Flame-of-the-Forest** (*Delonix regia*), from Madagascar, is often found covered in scarlet flowers in Singapore. The flowers can cover the whole tree for weeks at a time. The tree grows quite fast and can reach some 20 metres in height. The flowers are arranged in bunches, the prominent uppermost petal streaked crimson with a yellow centre. The broad crown makes this a good shade tree, although the big, buttress roots and frequent leaf-shedding make it unsuitable for narrow roadsides. It is a favourite street tree in tropical cities around the world.

Pacific Islands

● The **Cook Pine** (*Araucaria columnaris*) may grow up to 70 metres tall in its native Australia and Oceania. It is one of the few needle-leaved trees that thrives in the tropics. It was "discovered" by Captain James Cook in New Caledonia.

● The **Fiji Fan Palm** (*Pritchardia pacifica*), from the Pacific Islands, is thought to have originated in Tonga, despite its common name. It grows well in full sun, given adequate watering and fertiliser. It can grow up to 10 metres in height, producing large fan-shaped leaves. Palm trees are unlike other trees, in that they usually do not form branches. Most palms grow as a single shoot from the base, with a crown of large leaves, known as fronds, surrounding the shoot. When the central shoot dies, the palm is unable to produce any more leaves and will die soon after. That is why palms are found mostly in the tropics – a hard winter frost would easily kill off the central shoot, which is the palm tree's only means of making new leaves.

Australia

● The **Golden Myrtle** (*Xanthostemon chrysanthus*) grows well in full sun and can tolerate dry conditions. Its compactness, drought tolerance and low maintenance requirements make it a popular choice for road verges. Growing to 12–20 metres in height, it produces bright-yellow flowers arranged in large, dense, rounded, fluffy heads. The tree comes alive with the buzzing of bees when in bloom. The yellow contrasts well with the dark green of the leaves. It was introduced to Singapore in 1982 from Cairns, Australia.

● The **Bottlebrush Tree** (*Callistemon citrinus*), from New South Wales and Queensland, Australia, can grow up to 8 metres tall. The bright-red flower spikes are 4–10 cm long and give the tree its common name. They can last several weeks. The "bottlebrush" is actually a collection of thousands of tiny red flowers, with each bristle being the stamen of an individual flower.

"I was brought up on Fort Canning Hill in the days when it was a military camp, during the mid-1960s. Even today, when I visit the park on business, I can find myself lost in memories. It's a wonderful, forested area, a gem right in the heart of the city, steeped in history, not just for me personally, but for Singapore as well."

Kong Yit San, assistant chief executive officer, Parks Management and Lifestyle Cluster, NParks

Fort Canning

At Fort Canning Hill (once known as Bukit Larangan – Malay for Forbidden Hill – and later Government Hill), nature flourishes alongside history.

In the 1980s, archaeologists found ancient Chinese coins and ceramics here. Earlier, during construction of the Fort Canning Service Reservoir in the mid-1920s, the British found a cache of gold ornaments, now displayed in the National Museum of Singapore. These and other artefacts, linked to the Malay kingdoms of old, have enriched our understanding of Singapore's history before the time of Sir Stamford Raffles, supporting the account in the Malay Annals of a royal palace on the site.

At different times, the hill was the site of Raffles' bungalow, Singapore's first Botanical and Experimental Gardens, a Christian cemetery, a 19th-century fort, a World War II command centre and a post-war base for the British and Singapore armies. Raffles loved it so much that he asked to be buried here, his remains to be mixed with the ashes of the Malay kings (he was, in fact, buried in his home country). In 1859, some 68-pounder guns were among the artillery installed at the fort. One was fired at 5.00 a.m. every morning, signalling the start of the day. Until 1896, the guns also served as the city's fire alarm. The original

fort, demolished in 1907, never saw military action.

In 1972, the greenery around Fort Canning became known as Central Park, when the land occupied by the military merged with King George V Park, created before World War II. It took on the name Fort Canning Park in 1981. In the late 1980s and early 1990s, a master plan transformed the park into what we see today, highlighting its importance to the Singapore story. Today, underground military installations dating from World War II and other historical traces, as well as the outdoor performance space, attract thousands of visitors.

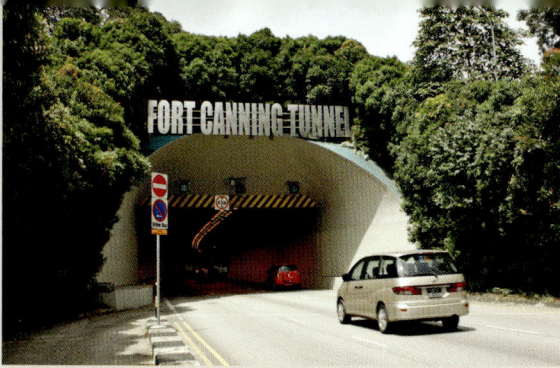

Tunnelling with minimum disturbance to trees

Construction of the Fort Canning Tunnel, linking the new downtown to the Orchard area, threatened some 22 mature trees in Fort Canning Park. So, rather than use the cheaper cut-and-cover method throughout, the Land Transport Authority (LTA) had special equipment brought in to bore a 180-metre section of the tunnel without disturbing the parkland above. The tunnel was opened in January 2007.

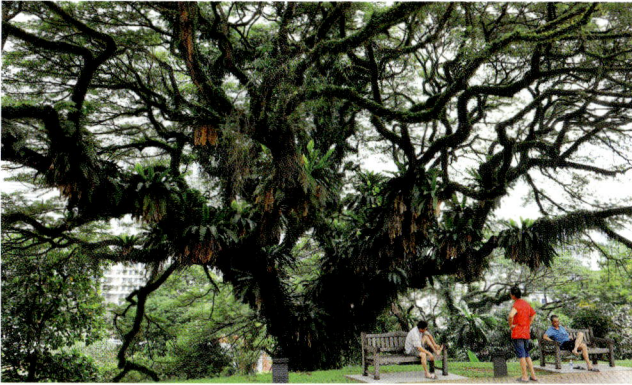

● Opposite: Rain Trees (*Samanea saman*) and strangler figs (*Ficus* sp.) flank the pathway leading to one of the Gothic Gates in Fort Canning Park.

● Above left: Relaxing under a Rain Tree.

● Above right: The stairway leading from Hill Street up to the flagstaff, shortly after redevelopment when the greenery was still immature.

● Below: Fort Canning Green, a popular venue for arts performances.

A softer touch

All over the world, cities built mostly of concrete have become drab. Apart from being ugly, bare concrete reflects glare and heat. Singapore has worked hard to avoid these effects – to create pleasant surroundings for living and working – ever since its transformation into a "clean and green city" became a national objective in the 1960s.

The surfaces of concrete structures were roughened and given a coat of stucco. Climbing plants became widespread, and today, Climbing Fig (*Ficus pumila*) softens the look of concrete all over Singapore. During construction of the East Coast Parkway and Pan Island Expressway, the then prime minister Lee Kuan Yew insisted that greenery should cover the ugly crash barriers.

Lee drew attention to the spaces under flyovers, the "cavernous depths", as he described them. With little light and moisture, plants do not grow well there. The GCAC considered artificial light, but rejected it on energy grounds. There was even a plan to use mirrors to reflect natural light! This was rejected as being too distracting to drivers. One creative solution was to split flyovers and elevated highways into two carriageways with a 1.5-metre gap between them, to let light and rain through.

Watering has always been an operational challenge. In the early days, some unexpected problems arose. On the Keppel Viaduct and later in Orchard Road, automatic sprinklers were tried. As P. Teva Raj, now director of industry and the Centre for Urban Greenery and Ecology, recalls, people kept removing the sprinkler heads, so the system had to be abandoned.

The planting of Bougainvilleas along overhead bridges and flyovers was an important innovation. It made a big impression on visitors and still does. However, it does not come cheap. Planting troughs are heavy, so the design of bridges was adapted to provide for the weight, as well as for watering and drainage.

● Opposite, top: Along Telok Blangah Road underneath the elevated highway grow *Cratoxylum cochinchinense* and *Sterculia parviflora*, both native tree species frequently planted alongside busy roads.

● Opposite, below: Stag's Horn Fern (*Platycerium coronarium*), planted on a roadside tree.

● Above: Bougainvillea softens the outlines of the pedestrian bridges across Eu Tong Sen Street and New Bridge Road in Chinatown.

Sprucing up

From the late 1970s to the mid-1980s, streetscape planting was part of an organised effort to "spruce up" Singapore. Hundreds of sites were planted with greenery. The authorities removed illegal dwellings and hawker stalls, and insisted on the repair of dilapidated buildings. Plants screened off slums in areas such as Rochor Road and Ophir Road. Greenery helped hide illegal squatter colonies, regarded as eyesores, from the eyes of visitors travelling from Paya Lebar Airport along Airport Road. Similar measures followed later along the road from Changi Airport and other important routes.

Orchids return to the roadside

Some 227 species of native wild orchids have been recorded in Singapore; 182 are now extinct, 39 are critically endangered or rare, and only four are common. NParks has reintroduced five of the endangered or locally extinct species. They are mostly epiphytes – they grow on other plants without feeding on them – and are being planted on trees in parks or by roadsides. One of them, *Grammatophyllum speciosum* (above), is the largest orchid in the world. A mature plant can weigh over a tonne. Its markings give it the common name Tiger Orchid. By 2013, over 800 Tiger Orchid plants have been planted around Singapore. They bloomed in that year on the roadside for the first time and could be seen along Holland Road, the East Coast Parkway and in East Coast Park. The flowers last about two weeks.

First impressions count

After only 20 years of operation, air traffic outgrew the international airport at Paya Lebar. Singapore needed a new airport, and it was to be at Changi. When it opened in 1981, some 6,500 plants decorated the inside of the terminal building (which later became Terminal 1). Outdoors, no fewer than 15,754 trees and 76,472 shrubs were planted around the expressway, ramps, carparks and other concrete structures.

The then prime minister, Lee Kuan Yew, was aware of Changi's importance to Singapore's image abroad. In 1988, he asked for the planting of "palm trees, to distinguish us from other places, because the airport building is not distinctive in shape…These palms will make it known that this is the Tropics, the equator, the exotic East."

The emphasis on greenery at the airport has continued ever since. More recently, travellers could enjoy a cactus garden, fern garden, orchid garden and sunflower garden. A refurbishment of the Terminal 1 building, completed in 2012, introduced decorative features inside the building reminiscent of Rain Trees and Tembusu trees. Terminal 3 features a display of vertical greenery five storeys high and 300 metres wide, containing 30,000 plants.

Because of its need for plants right from the day it opened, Changi set up its own nursery. Today, the nursery, with its own team of horticulturists, grows 300 species of plants, 40 palm species and 30 tree species.

● Below: Rain Trees, Bougainvillea bushes and *Bismarckia* palms line the road to Changi Airport.

● Opposite left: The view over the orchid garden pond at Terminal 2, Changi Airport.

● Opposite right: A green wall makes a strong visual impact on arriving passengers in the Terminal 3 baggage area.

"One day, after the airport was open, Dr Goh Keng Swee went to meet the Taiwanese ex-minister of defence, who remarked on the vegetation beside the ECP. 'How come all the trees are already growing?' the visitor asked. 'Good coordination,' responded Dr Goh."

Wong Yew Kwan, former commissioner of Parks and Recreation, Parks and Recreation Department

The view from above, then and now

More than half the area occupied by Changi Airport today is reclaimed land. The soil on reclaimed land is mostly sand, dredged from the sea, and lacks nutrients. Plants need a layer of organic material. Before the airport opened, much of the land beside the newly built ECP was covered in rough scrub and would have given a poor impression to incoming air travellers. The PRD officers acquired a huge quantity of seeds, put down a layer of topsoil and scattered the seeds on top. By the time the airport opened, greenery was growing in profusion. The contrasting views of the ECP above show young and mature Rain Trees respectively.

Creating homes

In 1960, Singapore had a serious housing problem. The government set up the Housing and Development Board (HDB) to tackle it. Within a few decades, HDB had transformed life in Singapore. At the outset, overcrowding and unsanitary conditions were rife. Just a few decades later, 83 percent of Singaporeans were living in well-designed and well-maintained public housing estates.

Vegetation was part of the solution. In the early days, while the architecture of estates was basic by modern standards, trees were planted between blocks and in carparks. To let air and water get to tree roots, greened strips and slotted paving were part of carpark designs (some early paving designs had to be modified, as ladies with high heels found them difficult to walk on!).

HDB's early towns had town parks. But from the 1980s, HDB's new town model included a refined hierarchy of parks and green spaces, resulting in each precinct having its own greens. With the upgrading programmes that began in 1989, older estates received a facelift with more landscaping, playgrounds and fitness corners. Open carparks were consolidated into multistorey carparks to free up space for more amenities. In younger towns, smaller common greens replaced neighbourhood parks, and these are conveniently located within walking distance of the heart of each housing estate.

HDB developed its own pool of horticulturists but continued to consult NParks on tree maintenance. For a time, it even ran its own nursery. However, as manpower became more expensive, HDB found that it was more economical to buy the wide variety of plants imported by the nurseries.

- **Opposite:** Greenery enhances an HDB estate at Pasir Ris Street 51.

- **Above left:** Semi-mature trees planted along roads and on top of a carpark form a seamless canopy when viewed from a flat in Punggol Oceanus estate.

- **Top right:** Palms and riparian plants by the banks of a landscaped drain create the impression of a forest stream in Bukit Panjang estate.

- **Above right:** Plants of various forms and colours growing on the carpark roof at Treelodge@Punggol help soften the concrete edges.

Gardening at high altitude

In 1992, the PRD and NParks jointly put on an exhibition to open the eyes of flat-dwellers to the potential for "skyrise gardens". A series of plant-identification cards (left) appeared at the same time. A useful little book, *Skyrise Gardening in Highrise Homes*, was first released in 1995 and is still selling nearly two decades later. It shows how to garden indoors, and that it is quite possible to grow foliage or flowering plants, fruits, vegetables and herbs in a flat. At the same time, HDB tries to make sure that gardeners, for all their enthusiasm, do not annoy their neighbours by spreading dirt in public areas or creating an obstruction. It points to the risks posed by pots on outside walls and windowsills, which can become "killer litter" if they fall off. Many would-be outdoor gardeners find it helpful to join a community gardening group and make new friends at the same time.

A new industrial landscape

In 1961, when Singapore's population was nearing two million and growing, it was calculated that over 80,000 new jobs would be needed over the next four years. Utilities such as water, gas and electricity supplies; communication links like roads, rails and telephone lines; and social amenities like housing, community and welfare centres had to be built. The word "industrialisation" was on every policymaker's lips. One of the first tasks of the new Economic Development Board (EDB) was to set up an industrial estate at Jurong, where Singapore's first heavy industrial plant, National Iron and Steel Mills, started operations in January 1964.

When the task of developing and managing the industrial estates became too large, the Jurong Town Corporation (now known as JTC Corporation) was established in 1968 to take over this role. Today, JTC reconciles business considerations with clean and pleasant surroundings, both in new projects and in the redevelopment of older estates, such as Tukang and Tanjong Kling. New trees have been planted to replace those lost. At the Seletar Aerospace Park, in the north of Singapore, native plants and other species are being planted along the roadsides to blend in with the landscape of the area. As

elsewhere on land managed by JTC, tenants follow JTC's landscaping guidelines.

For example, "one-north" is home to three research clusters, identified as biomedical sciences, infocomm technology and media industries. The research hub is master-planned to contain 16 hectares of green space located at one-north Park, a central green spine that connects the clusters together. The leafy colonial-era housing at nearby Rochester Park, Nepal Park and Wessex Estate has been incorporated into JTC's plans. Just as Singapore has matured, its industrial landscape has changed too.

"Quality of life is important for investors because you're not talking just about the factory. The machines don't care about fresh air. But the people who come to work, and their families, do worry about the environment. They want clean air; they want pleasant surroundings, and greenery is an important part of living in a happy place. It's very important for all Singaporeans."

Lee Hsien Loong, prime minister

● Opposite: The International Business Park at Jurong East incorporates landscaped grounds and water features to provide a conducive working environment for business executives and workers.

● Above and below: JTC has incorporated indoor greenery and sky gardens into its building, Fusionopolis, at one-north.

Assessing the impact

Environmental sustainability is critical to preserving our natural environment and creating a sustainable industrial landscape for future generations. Manohar Khiatani, CEO of JTC, explained, "As the largest industrial property developer in Singapore, we embrace environmental sustainability and have implemented numerous business initiatives to reduce the adverse impact of industrial activities within our estates on the environment. We also actively seek to combine ongoing innovations in environmental technology with environmentally-sound business practices. For example, to improve energy efficiency, we consider deploying green technologies in our developments."

Environmental assessments were conducted for the construction of Pasir Panjang Ferry Terminal, the coastal protection and mangrove enhancement project at Pulau Tekong and many other projects. Dr Lena Chan, director of the National Biodiversity Centre, NParks, explained, "It is important to conserve our native biodiversity, because it is our natural heritage, and for the essential ecosystem services it provides. Through the assessment process, the survival of our biodiversity is taken into account, within Singapore as well as in the region."

Bukit Timah Nature Reserve

What did the interior of Singapore look like two centuries ago? The answer can be found on Singapore's tallest hill, in the Bukit Timah Nature Reserve. Here, the giant trees in their natural setting are remnants of the forest that once covered most of the island. The name "Bukit Timah" means literally "tin hill" in Malay, but one theory suggests that "Timah" is a corruption of "temak", a local name for a subgroup of the *Shorea* genus of Dipterocarp trees.

According to one 19th-century record, after John Crawfurd, one of Singapore's first colonial administrators, climbed the hill in 1825, he reported that the surrounding forest was so dense that it would have been almost as easy to sail to Calcutta as to reach Bukit Timah! The access road to the summit was not built until 1843. The reserve has been protected by law

since 1939, when encroachment by granite quarrying nearby became a serious threat.

By the 1990s, the public was becoming very interested in nature. Visitor numbers at Bukit Timah Nature Reserve were rising dramatically. In 1990, the reserve saw some 88,000 visits. By 1998, the number was around 400,000! There was a risk that the reserve could become a victim of its own success, that the public might, literally, "love it to death".

In 1991, the newly formed NParks conducted a visitor survey and, on the basis of feedback, began developing facilities to make the reserve a more pleasant place to visit and at the same time, reduce the impact of visitors. Over time, the network of trails had penetrated sensitive areas of the forest, leading to attrition in animal and plant numbers,

damage of the ground surface and erosion problems. So the nature reserve staff closed some of them off and upgraded the main trails.

More relief came as visitor facilities in the Central Catchment Nature Reserve were improved and areas close to the Bukit Timah Nature Reserve were designated as nature areas to absorb some of the pressure.

In 2011, the Bukit Timah Nature Reserve was declared Singapore's second ASEAN Heritage Park (the first being Sungei Buloh Wetland Reserve).

● **Above:** The visitor centre at Bukit Timah Nature Reserve.

● **Opposite:** A Seraya tree (*Shorea curtisii*), the dominant canopy tree species in Bukit Timah Nature Reserve.

An aerial view showing Bukit Timah Nature Reserve at bottom right and Hindhede Quarry in the centre.

"I go to Bukit Timah quite regularly. You're so close by and yet in such a radically different environment. You can imagine what it was like before Singapore got developed."

Lee Hsien Loong, prime minister

Nature is complex

"Primitive forest in Malaya is highly complex. It consists of some hundreds of kinds of trees; also lesser plants of many kinds which grow in the shade of the trees, orchids and ferns which grow in varying degrees of exposure on the trunks and branches of the trees and a great variety of woody climbers…which use the trees as supports.

The forest has an internal climate which is quite different from that outside. The controlling factor is the shade cast by the tallest trees, to which is added the further shade of the lesser trees, so that at ground level there is a weak light, little movement of air and consequently a high humidity.

Furthermore, a layer of dead leaves accumulates on the ground, and by their gradual decay these feed the roots of the living plants. The decaying mass also forms a sponge which absorbs rain water, checks erosion, and by remaining moist, maintains the humidity of the air near the ground…

A complex forest, such as I have described, is self-perpetuating if it is undisturbed; but its existence depends on the maintenance of the internal climate, and that depends on sufficient tree cover…"

R.E. Holttum, former director of the Singapore Botanic Gardens, writing in The Straits Times *(1950)*

Are there tigers?

When the pioneering naturalist Alfred Russel Wallace studied wildlife at Bukit Timah in the mid-19th century, he heard the roar of tigers (*Panthera tigris*). He must have had strong nerves – tigers were, he reported, killing one person a day on average. Most of the victims were workers on gambier and pepper plantations. Singapore's last wild tiger was killed in the Choa Chu Kang area in the 1930s. In 1997, a tiger was suspected of leaving large paw prints on Pulau Ubin, but the animal concerned turned out to be a large dog.

● **Opposite:** Visitors enjoying an early morning walk.

● **Top right:** The Plantain Squirrel (*Callosciurus notatus singapurensis*) is a common sight in nature areas, parks and housing estates.

● **Centre far right:** The Common Sun Skink (*Eutropis multifasciatus*) lives in wooded areas, mangroves and parkland.

● **Centre right:** The Lesser Dog-faced Fruit Bat (*Cynopterus brachyotis*) lives in trees and buildings.

Replacing the missing link

In 1986, construction of the Bukit Timah Expressway cut the Bukit Timah Nature Reserve off from the Central Catchment Nature Reserve. The two habitats are now being reconnected. The LTA, working with NParks, started construction of the Eco-Link, a green bridge across the highway, in 2011.

For the designers of the Eco-Link, one of the main challenges is ensuring that it provides the right habitat for the wildlife. They also need to use physical barriers to guide animals towards and across the link. Over time, it is hoped the wildlife in both nature reserves will benefit. NParks will monitor the Eco-Link's success by animal tracking.

Istana

The Istana is open to the public about five times a year. As visitors walk up through the Centre Gate (marking the boundary of the grounds in times past), it is easy to see why foreign VIPs are often impressed. The house commands a sweeping view over the Ceremonial Plaza, where visiting heads of state inspect a guard of honour, down towards the front lawn. There are some 8,000 trees in the gardens, representing more than 100 species. Some are very old – a few Tembusus are thought to be older than the Istana building itself, which dates from 1869. The land occupied by the gardens was originally a nutmeg plantation.

The range of plants in the gardens today reflects, in part, the personal commitment of Lee Kuan Yew. He encouraged the gardeners to try anything that might grow well in local conditions, bringing a variety of shapes and colours to the gardens. He was supported by Mrs Lee, whose love of fragrant plants made a big impression on the garden staff.

"Just imagine what the Tembusu in the Istana gardens has witnessed over more than 150 years. It's here because somebody in the past protected it and left it with us. My role now is to protect it for the benefit of future generations. At the same time, we have to plant new trees because the old trees may die. You can never sit back."

Wong Tuan Wah, senior curator, Istana

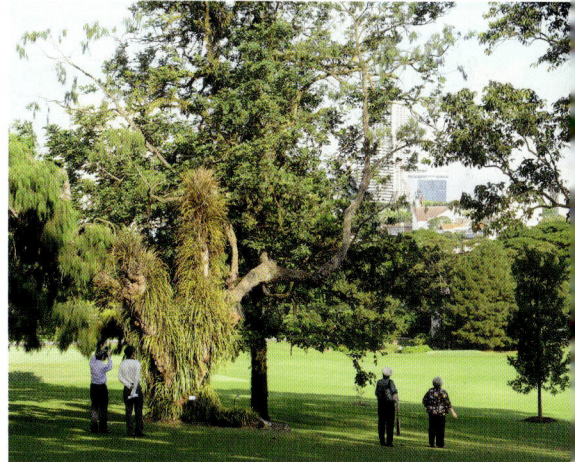

● The Istana Government House (as it was then called) and the surrounding gardens, in the 1870s (opposite, top) and today (this page, top).

● Opposite, below: The view over the Istana Lawn towards the city.

● Above: A Weeping Ru (*Gymnostoma rumphianum*) on the Istana grounds.

Birds of the Istana

Around 2008, two pairs of Oriental Pied Hornbills (*Anthracoceros albirostris convexus*) made their home in the Istana grounds and have since produced several chicks. The young do not live with the parents for long. As curator Koh Soon Kiong explained, "At the start of their annual breeding cycle, they chase previous offspring out of their territory. We paired up one of the male offspring with a female from Pulau Ubin and took them to Sungei Buloh. They have started breeding there." Koh and his colleagues also wanted to increase the population of Oriental Magpie Robins (*Copsychus saularis*). Having set up nestboxes made out of old biscuit tins, the curators introduced a number to the Istana gardens, and a few pairs regularly now come to the feeding tables for a tasty meal of worms.

2

BREAKING NEW GROUND, REDISCOVERING ROOTS

With the arrival of the 1990s, radical changes came to the parks and nature reserves. The focus shifted to meeting the aspirations of people through better recreational facilities, greater diversity of parks and more imaginative landscaping. A master plan for the Singapore Botanic Gardens heralded not only expansion and new features but also a return to its core mission. Conservation became a focus in its own right. Driving these changes was a new organisation – NParks.

● With a new generation of park visitors and nature lovers, parks have to evolve in both design and the type and range of amenities provided.

New thinking needed

In the 1990s, park users were becoming better informed, more cosmopolitan and more sophisticated. Dr Kiat Tan, who headed the newly formed NParks from 1990 and later became CEO of Gardens by the Bay, recalled, "In the early days, land was allocated to parks because it was not economically useful. Parks tended to be far away from where people lived, patches of land that would not get in the way of later development. With the building of new towns, and increasing cooperation between government agencies, that all changed – but the key underlying factor was public engagement."

The goal of park managers in the 1970s and 1980s was to fill the parks with healthy-looking trees and immaculate landscaping. Understandably, managers preferred trees they were familiar with. There was little incentive to take risks and introduce new things. The "clients" were the people who provided the budget, rather than the public. Some politicians and civil servants took a sceptical view of parks in general and saw innovation as a waste of government money.

Changing the way parks were managed required organisational change. In January 1983, when Dr Tan joined the PRD as assistant commissioner in charge of research and the advisory programme, there was little communication between planning and research and the maintenance division. A silo mentality had arisen. The early staff of the PRD, many of them with an agricultural or plantation background, were guided by directives from the GCAC. The result was plenty of shade trees, but also a lack of variety, of which even the prime minister became aware. This is what led to a new emphasis on colour.

The need to try out a different kind of organisation led to the formation of the National Parks Board in 1990.

"In the past, many decisions about greenery were 'top down'. That will be different in the future. We must be in tune with what people want and how lifestyle expectations are changing. To engage the community is absolutely critical. Future planning must reflect that kind of thinking."
Kenneth Er, assistant chief executive officer (special projects), NParks

Dr Tan wanted, in particular, to restore professional skills to their rightful place and thus reinstate Singapore Botanic Gardens in its professional role. This, in his opinion, was only achievable outside the PRD. The permanent secretary of MND agreed to the formation of a new statutory board. As Dr Tan was already secretary to the Nature Reserves Board, his new portfolio included the reserves and also Fort Canning Park as CEO. In 1996, NParks took over the entire park management role of the PRD.

After the creation of NParks in the 1990s, the style of park management changed, and increased attention was paid to the wishes of the public.

The first National Parks Board logo, designed by Lionel Low Wai Chin, was the winning entry in a public competition.

The present-day logo, dating from 1996, was designed by Eng Siak Loy and reflects the expansion of NParks' role at that time. Eng also designed the new gates at the Tanglin entrance of the Singapore Botanic Gardens.

A time of innovation

Soon after the formation of NParks in 1990, master plans were drawn up for the reinvigoration of the Singapore Botanic Gardens and Fort Canning Park. "Programming" was introduced, involving the public and students in events and activities. Close attention was paid to visitor numbers and public feedback. More facilities were provided, such as toilets, benches, shelters and lighting.

The nature reserves became more accessible, without undermining their basic function. In 1990, the operating budget for the whole of Bukit Timah Nature Reserve, managed by just a small team of rangers, was less than $100,000. As people's perception of the reserves changed, and they developed a passionate interest in nature, visitor numbers grew and budget increases naturally followed.

The innovative approach of the new NParks was successful. In 1996, the rest of the then PRD was brought under the wing of NParks, thus creating the organisation we see today. Park planners looked to see where parks would have the greatest public support. Park managers were asked to make parks more attractive to people. Managers had to be more creative. Even if some plants died, it was treated as a learning experience.

The researchers were no longer cut off from the park managers. Each park management team was to have access to a full range of knowledge and skills: botany, horticulture, park management and leisure management.

Green spaces were formerly provided by the government in a "top down" manner as a form of infrastructure. Now planners paid more attention to the needs of the public, who were beginning to gain some purchase on their own environment. As Dr Kiat Tan put it, "NParks stewardship can only be sustained if the public takes over the reins."

"We put the 'recreation' back into 'Parks and Recreation'."

Dr Kiat Tan, chief executive officer, Gardens by the Bay,
and former CEO, NParks

● This page: Kallang Riverside Park during construction of the National Stadium in the early 1970s (above) and today (right).

● Below: Innovative play equipment at West Coast Park.

Gardeners for the nation

The NParks mission is to

- establish world-class gardens
- rejuvenate urban parks and enliven the streetscape
- optimise urban spaces and infrastructure for greenery and recreation
- enrich biodiversity in the urban environment
- ensure competencies of the landscape and horticulture industry
- engage and inspire communities to co-create a greener Singapore

"I couldn't really appreciate what NParks was doing until I went outside Singapore – and I could see what Singapore might have looked like without coordinated efforts to transform it into a garden city."

Marcus Chua, ecology researcher

NParks officers have to work as a team to maintain parks and the health of the plants within.

How much land does NParks manage?

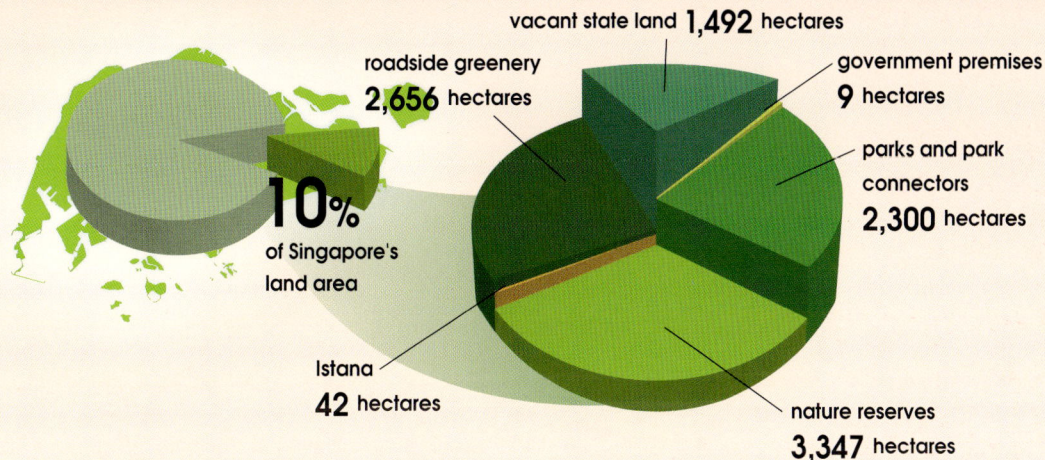

vacant state land **1,492** hectares

government premises **9** hectares

roadside greenery **2,656** hectares

parks and park connectors **2,300** hectares

10% of Singapore's land area

Istana **42** hectares

nature reserves **3,347** hectares

How many parks does NParks manage?

254 neighbourhood parks

62 regional parks

4 nature reserves

46 park connectors (over 200 km)

NParks electronically monitors around **1.4 million** trees in parks and along roads.

Who does the work?

NParks staff strength: **957**

Contractors' workforce: approx. **2,500**

How much does it cost to keep up the parks?

Annual expenditure on park maintenance and improvement, excluding the cost of NParks staff: **$62 million**

*All figures are from March 2012.

"There are opportunities for Singaporeans to gain professional qualifications and grow with the industry. For example, more and more greenery is being provided in private developments. You need supervisors to maintain it who understand all the technicalities. The industry is growing and creating hundreds of new jobs that require skill and knowledge. Singaporeans could do them."

P. Teva Raj, director of industry and the Centre for Urban Greenery and Ecology

The importance of expertise

In 1972, the Singapore Botanic Gardens set up a School of Ornamental Horticulture, offering full and part-time diploma courses. Initially, it trained around 10 people annually to work on the greening of Singapore. Most of them went on to work in the PRD; some went to other statutory boards and to private nurseries. The school had no research role. Later came the School of Horticulture, which awarded diplomas and trade certificates in tropical horticulture, and landscape management and design.

Ideally, these diplomas would have been reflected in salaries. In fact, they were not seen as equivalent to tertiary institution qualifications, and so training shifted to the tertiary institutions – Ngee Ann Polytechnic, in particular.

More recently, NParks revived the idea of having its own training institution. As a result, with input from the Singapore Workforce Development Agency, CUGE (Centre for Urban Greenery and Ecology) opened in 2007. It provides skills training and professional certification for the landscaping industry in general. In its first five years, it trained more than 6,000 locals. The courses, which comply with Singapore's Workforce Skills Qualification (WSQ) scheme, are modular and designed for people already in employment. They focus on specifics such as turf management, irrigation, potting, composting and mulching. The range is wide, from playground management to water-sensitive urban design. Because of the small size of the industry in Singapore, those requiring a bachelor-level course in horticulture have to look overseas.

CUGE also carries out research. Its programmes cover the performance of plants, such as different kinds of grass, in urban conditions; the integration of greenery, including skyrise greenery, into the built environment; and urban ecology, including pest and disease management.

The landscaping industry itself plays a central role in the development of management and technical skills, and the maintenance of standards. Its efforts are coordinated through the Landscape Industry Association (Singapore) (LIAS).

Left to right: The PiCUS tomograph uses sound waves to detect potential pockets of decay in a tree; assessing decay in a tree, using a resistograph, which measures physical resistance to drilling; CUGE trains a student in the art of tree pruning.

Below: An instructor shows how to use a chainsaw safely.

"I took a diploma course in horticulture and landscape management at Ngee Ann Polytechnic. Unfortunately, if you want to take your training to degree level you have to go overseas. I wish gardening was more attractive to young people as a career."

Felicia Wei, site supervisor, Prince's Landscape & Construction

The nursery

In the 1970s, the PRD ran several nurseries dispersed across Singapore, but soon they were consolidated into just two. One of these, the Changi nursery, had to make way for the airport. Now that many plants are available from private nurseries, NParks has just one major nursery, and that is at Pasir Panjang.

NParks officers pass their plant requests to the nursery online. According to Lilian Kwok, head of nursery management, demand has soared – compared to 100,000 plants annually just a few years ago, the number for 2011 was around 150,000. The 2012 total was around twice

that – many of which were used for landscaping at the new Gardens by the Bay. The top sellers are the native plants that are not easily available in other nurseries. Nursery staff have a weekly programme to collect seeds of native plants from the forests and parkland across Singapore.

The nursery is about the size of 17 football pitches. It employs a team of seven experts on horticulture, arboriculture, agriculture and entomology (the last of these is key to a knowledge of insect pests). They each focus on one or more plant categories such as palms, aquatic plants, plants that attract

butterflies or birds, native trees and shrubs, ornamental trees and shrubs (many of which are introduced species) and salt-tolerant trees for planting near the seashore. One area is set aside for use by the National Orchid Garden.

The staff at the nursery love their plants. Some have carried on working well beyond the normal retirement age of 62. Much of their knowledge is the kind that can only be gained through hands-on experience. They often advise landscape architects, whose design skills are not always fully matched by detailed plant knowledge.

● Above: Growing plants from cuttings is much faster than growing them from seeds. Pictured here (from left to right) are a root ball; a nursery staff member watering a seedling of *Ardisia elliptica*, a small native tree; tending to cuttings of *Rhodomyrtus tomentosa*, a native shrub; and planting cuttings from a Tampines Tree (*Streblus elongatus*).

● Below: A general view over the NParks nursery at Pasir Panjang, showing a variety of trees, palms, shrubs and aquatic plants.

● Top right: Young tree saplings and shrubs have to be kept under shade, as direct sunlight tends to dry them out.

● Right: Shrubs and trees awaiting collection for planting in parks and alongside roads.

Singapore Botanic Gardens

Singapore Botanic Gardens attracts over four million visits annually. It is respected worldwide. The scientific knowledge accumulated here is an important resource for the nation. The beautifully landscaped gardens are a "green lung" for the people.

The land occupied by the Gardens is a former gambier plantation. It was opened in 1859 as a pleasure garden by the Agri-Horticultural Society, which organised flower shows and horticultural fetes. By 1875, it was under government control and became part of a worldwide network of gardens linked to the Royal Botanic Gardens at Kew, London, promoting research into economic and medicinal plants. Under its first director Henry N. Ridley, the Singapore Botanic Gardens promoted rubber as a commercial crop before the motor industry took off worldwide. The timing was just right – all those cars needed tyres. Oil palm (*Elaeis guineensis*) was another successful crop. Singapore's economy grew.

In the 1970s, when the campaign to turn Singapore into a garden city was in full swing, the focus of work at the Gardens moved away from research towards areas more immediately useful to the greening campaign, such as tissue culture, pests, diseases and the training of PRD staff. A revival of the research tradition came with

● Above: Families enjoying a performance at the Shaw Foundation Symphony Stage.

● Left: The Tanglin Gate.

● Opposite: Swan Lake, the oldest man-made lake in Singapore.

The Bandstand, for generations a favourite spot for romantic rendezvous. The trees are Rain Trees (*Samanea saman*), showing an interesting yellow-leaf mutation.

The biggest orchid garden in Asia

Orchids have been a feature of the Singapore Botanic Gardens since the 1870s. Breeding started in 1925 under director R.E. Holttum. He experimented with propagation in test tubes, and in 1929, produced the Gardens' first hybrid, *Spathoglottis* 'Primrose' (*Spathoglottis aurea* x *Spathoglottis plicata*). In the following years, the orchid industry became a crucial one for Singapore.

In 1995, the National Orchid Garden, which took three years to build, replaced the former Orchid Enclosure (where the Ginger Garden is today). It contains over 1,000 species and 2,000 hybrids, with about 600 on display. More than half a million visitors each year enjoy the colourful blooms – yellow, cream and golden flowers to represent spring; pink and red for summer; red and purple for autumn; and white for winter. Singapore's national flower (inset) is an orchid, *Vanda* 'Miss Joaquim'.

● Opposite: The attractions at Singapore Botanic Gardens include the Eco-Lake (centre right) and the Evolution Garden (centre left).

● Opposite, top: Burkill Hall, built in 1866, commemorates the only father-son pair, Isaac and Humphrey Burkill, to hold the post of Director of Singapore Botanic Gardens.

A priceless collection

Over the years, many famous botanists have headed the Singapore Botanic Gardens. Their legacy includes the herbarium, founded in 1875. It is a "library" of plant material, containing around 650,000 specimens that are very valuable. Some 6,800 are "type" specimens, the original specimens from which plant species were first identified and given botanical names.

the formation of NParks in 1990, under the leadership of Dr Kiat Tan. He drove forward a new master plan for the development of the Gardens, having in mind not only the importance of deep professional knowledge, including research, but also the desires of the public, whose engagement he saw as essential to the Gardens' future.

Singapore Botanic Gardens has re-embraced its scientific and educational mission, with a focus on taxonomy, orchid-breeding and micropropagation. Current research includes the re-establishment of Singapore's native biodiversity. If plants now extinct in Singapore can be found growing elsewhere, it may be possible to reintroduce them.

Singapore Botanic Gardens changed dramatically in the 1990s. It gained new visitor facilities and attractions. The area around the Nassim Gate on Cluny Road was developed to make it easier to reach the magnificently landscaped Palm Valley, Symphony Lake, Ginger Garden, National Orchid Garden and Rain Forest (one of the few remnants of primary forest in Singapore).

The Tanglin Gate, on Holland Road, leads into the oldest part of the gardens, including Swan Lake. In this area, too, there are new facilities, such as the herbarium building and the Botany Centre, with a library open to the public.

In the newest part of the Gardens, near Bukit Timah Road, the focus is on entertaining and educational projects, designed to get people involved. A pond originally created to retain storm water became the Eco-Lake.

The Evolution Garden is a landscaped walk illustrating the origins of the world's plant life and its evolution into myriad complex forms well before the arrival of humans.

The Healing Garden showcases over 500 varieties of plants traditionally used medicinally in Southeast Asia. It is laid out thematically, relating to parts of the body such as the head, respiratory and reproductive systems.

The Fragrant Garden brings together a collection of plants appealing to the sense of smell.

The success of Singapore Botanic Gardens is due in no small part to the support of friends and partners who have contributed to important projects, such as the Orchid Conservation Fund, the Shaw Foundation Stage, Jacob Ballas Children's Garden and many more.

In future, the Gardens will be even more accessible to the public. In 2011, a new MRT station opened near the Bukit Timah gate, and the authorities have planned for another station to be sited near the Tanglin Gate.

Building for the future

The Jacob Ballas Children's Garden is a place where the under-12s can discover the wonders of plants through play and exploration. The garden attracted 400,000 visitors in the 12 months after its opening in 2007. The theme is "All life on Earth depends on plants". The garden shows how plants provide for our daily needs. A visit to the garden is fun and, at the same time, an enjoyable introduction to life sciences.

A day in the life of...

To maintain a city in a garden involves many kinds of expertise, brought together in a huge team effort. Creating a park is just the beginning. Looking after the plants and animals that live there, as well as maintaining and improving public facilities, is a never-ending task calling for endless commitment. The next six pages profile the jobs of just a few of the team members.

Koh Kar Yan, manager, coastal parks, is in charge of Sembawang Park. She takes great pride in the park, where she usually spends her mornings. There is a small office there with Internet connection so she is never out of touch for long. Tree inspections are an important part of her job. She must always be on the alert for problems. The park is one of a cluster of five coastal parks, and on average twice a month she travels to other parks, such as East Coast Park, to help with inspections there.

The park entered a redevelopment phase in 2011. Although some areas had to be hoarded up, she still has to make sure that the trees within the construction zone are being maintained properly. She inspects all the trees in the park regularly, following a systematic checklist. If there's a problem, she can call on one of her colleagues who specialises in arboriculture.

The most popular park playgrounds are those in the coastal parks. Kar Yan regularly checks the play equipment and all the other public facilities. A simple thing like a broken chain on a swing can be dangerous.

Kar Yan has to liaise with contractors all the time, whether they are working on the redevelopment or on routine maintenance. As a result, she has picked up many non-core skills. "I have learned to be a coordinator," she says. "When several parties are involved, representing NParks, contractors or other government agencies, progress can easily grind to a halt. It is my job to take the initiative and move things along."

Kar Yan has a degree in geography, which is handy for the manager of a coastal park, where issues such as climate change, sea levels and coastal erosion are very relevant. "It was the flora and fauna that first attracted me to NParks. The biggest challenge was learning the administrative systems, but I have learned to combine idealism with the realities of doing a job." As she has discovered, looking after a park these days requires management skills, and she spends about half her week at NParks headquarters, working on matters such as tender evaluations.

She still finds time to put her interest in biodiversity to good use. As a volunteer, she has been helping with a survey on fireflies at the Pasir Ris park mangroves – they are an "indicator species" reflecting the more general state of the ecosystem. She also enjoys interacting with people. "We have regular activities to educate park users about habitats such as the mangroves. I enjoy helping out and getting to meet the public."

New playgrounds do not have to be boring. Recently completed at Sembawang Park is a huge "battleship", the centrepiece of a new adventure playground. It is quite high and complex in design – the designers wanted to challenge the children, to get them to climb and explore. It echoes Sembawang's maritime connections – the shipyard nearby started life as an important naval base in the 1930s.

Yee Chung Yao is in charge of the park connector team, which oversees the maintenance and management of the network. Because the network is growing, he is also deeply involved in its planning and construction.

Chung Yao has to travel all over Singapore, visiting places where the Park Connector Network is being developed. He often accompanies NParks colleagues working in areas such as planning, development, streetscape management or park management. They have to figure out exactly how new connector stretches should be aligned. Sometimes a planned alignment might have to be shifted to a parallel track or the opposite bank of a waterway, to accommodate, say, a water pipe.

Chung Yao was trained in zoology, but his job covers more than plants and animals. While he is out and about he does spot checks, identifying defects and supervising maintenance. All the time, he is looking out for trees with low-hanging branches or roots affecting the connector surface. Many of his concerns are basic – general cleanliness, ensuring toilets are kept clean and rubbish bins regularly emptied.

Although he prefers to be in the open air, Chung Yao has to spend a bit of his working week back at NParks headquarters. There, one of his responsibilities is to manage public feedback. "The most common request is for more shade," he says. "Unfortunately, forest trees are not best suited for the connectors, although salt-tolerant species may be planted near the coasts." In more built-up surroundings, it is more effective to use ornamental shrubs, which bring colour to the scene.

Chung Yao meets representatives of the public, such as MPs and grassroots leaders, as much as he can – they provide valuable feedback. In addition, he says, "We have also conducted bicycle tours of the connectors for our partners in other agencies, so they can give their suggestions for enhancing the network, how to bring in more users, serve a bigger public, be they commuters or tourists." He and his team have created outreach "packages" for the public, with themes such as safe skating, or being a responsible park connector user. They have worked with sports organisers, including those of the Sundown Marathon, Singapore's first night marathon.

Boo Ghim Yew is an arborist working for Arbsolutions Asia, which gives expert advice to corporate or individual clients. He spends much of his time looking at development plans. For example, a developer planning a new condominium may need to comply with regulations as to how many trees he is allowed to fell and how many will be retained. Some trees might be affected by drains or piling. The issue is particularly sensitive in Tree Conservation Areas, where mature trees are protected by law. After familiarising himself with the case, it is Ghim Yew's job to inspect the site. He then returns to the office and writes a report, which forms part of the development plans that the client submits to NParks.

According to Ghim Yew, "These days, developers generally bring in arborists early in a project's life. This is much better than a few years ago, when NParks approval was often sought after all the other agencies, by which time it was difficult to change anything." While Ghim Yew's firm works for the client, he says, "I also represent the trees' interests. Our concern is the health of the trees, and we give objective advice. Clients are well advised to take our advice, as it speeds up the approval process."

Ronnie Tan is a landscape architect working for Stephen Caffyn Landscape Design, handling public and private projects, from creating private gardens to master-planning big, mixed-use developments in Singapore and abroad. He is also second vice president of the Singapore Institute of Landscape Architects (SILA). He was trained in New Zealand. "I went into landscape design because I was interested in design and had a love for nature and the outdoors. I like to see the results of my work. And this is a profession where I can help mitigate the effects of factors such as global warming."

Ronnie spends part of his time in the office, part of it in the offices of his clients and part on site. The key issues vary from project to project. "If I were working on, say, a neighbourhood park, our focus would probably be more oriented towards providing a leisure space with facilities for the residents. Other projects might be focused on environmental conservation."

Developments should not impact the natural environment more than necessary. "It is in everybody's interest to minimise the impact of development and conserve what is already there," Ronnie says, "but one also has to maximise the use of the site by stakeholders. Therefore we have to manage the different conflicting demands."

Ronnie's firm is applying its experience in countries overseas, such as China. Design principles are usually constant in these overseas projects, involving effective space planning, aesthetics and conservation of the natural environment. What varies are the unique cultural, ecological and climatic conditions of each site. Projects in Singapore need plenty of shade to mitigate the hot climatic conditions, whereas in colder temperate countries, open spaces allow people to soak up more of the sun.

Dr Nura Abdul Karim heads the team that keeps records on all the plants in the Singapore Botanic Gardens. When a plant comes in, she and her colleagues at NParks have to verify its source and scientific name, and assign to it a unique identification number. Accurate labelling is essential. Throughout the life of the plant, she can monitor where it has been placed, where it is moved to – and when it dies. Dr Nura explained, "Sometimes I have to ask for taxonomic information from the herbarium; sometimes I have to send off emails to researchers overseas who I know are concentrating on certain families – it's a bit like being a detective!" The result is a massive database useful to researchers and conservationists, particularly those interested in rare plants. Many plants are acquired by means of seed exchanges with botanic gardens overseas, which maintain lists of available plant materials.

Dr Nura answers enquiries not only from specialist researchers but also from the general public. "Many questions are basically, 'What is this plant and how do I grow it?' People ask about flowering and fruiting seasons. They ask if we have a specimen of the plant in question. Edible plants are a particular favourite."

Maintaining the database is a big job. Dr Nura tries to ensure that each area of the Gardens is checked at least once a year. "My biggest challenge so far has been to survey the Singapore Botanic Gardens' tropical rainforest. We started in 2009, spending half a day a week, involving an arborist, the lawn manager, two of my staff and one of the herbarium research officers who is good at field identification. We had to create a detailed map and identify and record the very tall forest trees."

Dr Nura is unwilling to pick out a favourite plant. "I enjoy them all," she said. "Every single plant is unique, has its own beauty and interesting facts. Even weeds are interesting."

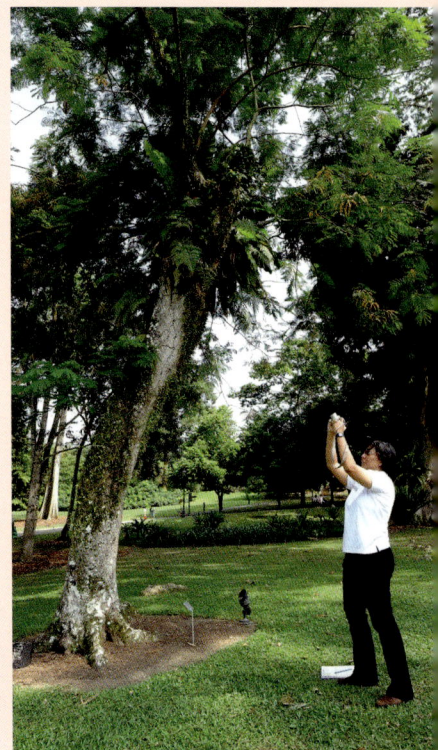

Clayton Lee is one of NParks' certified arborists. He could be described as a consultant tree doctor. Every day he receives a list of trees that need examining. It comes to him at NParks headquarters via his section head, often with a photograph uploaded into the NParks system by the officer on the spot. The managers perform regular inspections on several hundred trees a month. When they see a possible problem, they access NParks' geographic information system and file a report, identifying the tree, its location and the nature of the defect. They describe the tree's general health and give information on the crown, branches, trunk and roots, as well as the site conditions and the potential impacts of any tree failure.

Clayton has to respond to every request for a site visit within a maximum of two weeks – less if the situation is urgent. The number of visits in a day varies. It depends on the location and nature of the problem. "We need a lot of information," Clayton says, "so that we will know what kind of equipment and manpower we will bring with us. We might need ladders, a lorry crane, or ropes and harness if we have to climb the tree ourselves for a closer look." The inspection team, which can number up to four people, is highly trained technically. They may need a resistograph – variations in drilling resistance at different points in the wood can indicate internal decay – or a sonic tomograph, which uses sound waves.

Having made his diagnosis, Clayton recommends a treatment. Sometimes he has to supervise it, for example, if sensitive pruning is needed. Contractors do the hands-on work. The local climate sometimes brings stormy conditions, including fierce gusts of wind. If widespread damage is feared, the whole of NParks is immediately put on the alert. Clayton then has to join a massive team effort, inspecting all the trees of the parks, streets and even the forests.

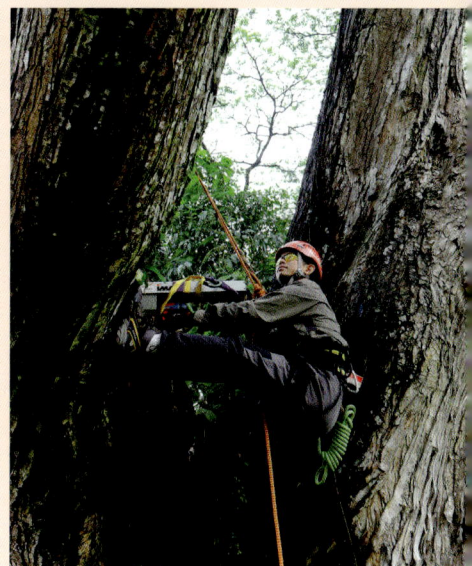

Has Clayton had any unexpected encounters in a tree? "I have met snakes, bees and even bats – occasionally a flying lizard. If you are working in the forest you might see a Colugo if you are very lucky."

Marcus Chua is a researcher specialising in ecology. Asked how he spends his day, Marcus replies, "The question should be: how do I spend the night?" As part of his postgraduate work at the National University of Singapore, he is studying leopard cats, which are nocturnal.

Marcus goes out at night with his survey team to the habitats where they think the cats might be. They follow pre-designated "transects", or patrol strips, so that trends can be identified on a consistent basis. Some members of the team work for NParks, some are students, some have volunteered through the Nature Society (Singapore). "Four pairs of eyes are better than one," Marcus says, "and you need a team for safety reasons, but we keep the number down to four so as not to scare the wildlife." Once a cat is spotted, the team plots its distance from the centre of the transect. Nocturnal animals have reflective eyes so that they can see better in the dark. "From a distance of up to 150 metres, we can see their 'eyeshine' – from the layers of reflective cells on their retina." The eyeshine colour helps them decide if the animal is a leopard cat, a domestic cat or maybe a wild boar. From regular observations, Marcus builds up a picture of the habitats the cats can be found in and their numbers. "There are more individuals than we thought – they are clearly breeding in Singapore. They are also doing well on Pulau Tekong and can be found on reclaimed land, a habitat where they have not been recorded before."

Marcus also uses camera traps, little cameras attached to trees, linked to motion or heat sensors. They will operate 24 hours a day, recording time and location. "We can see how many individuals there are. Sometimes there will be adults with kittens. We have discovered interesting data on other animals, including some that are very rare or not previously recorded in a particular area. An example is the Greater Mousedeer in the Western Catchment area – previously it was thought to occur only on Pulau Ubin. We have also photographed monitor lizards mating, something you don't often see. Sometimes we see humans doing surprising things."

Marcus's early interest in ecology was sparked by books and TV documentaries. He developed it at school and then at NUS, where there was a programme for life sciences (now called environmental biology).

Felicia Wei works for Prince's Landscape & Construction. She oversees teams of workers who do maintenance and project work and set up plant displays for major events. They are permanent employees, half of them Singaporeans – the government sets a limit on the number of foreign workers that can be employed by a landscape company.

Her day starts early. She has to brief her teams on the day's tasks and on safety before 8.30 a.m., when they head out to their worksites, kitted up with safety equipment. On site, there is no time to waste. Site supervisors go out with the workers. On complex jobs, the supervisors will stay to keep an eye on things; mostly they do the rounds, covering several teams during the day.

Regular tasks include watering, fertilising, grass-cutting and pruning. The workers pick up plant knowledge quickly. "We give them lessons in plant identification once a month. One definition of a 'weed' is a plant growing in the wrong place. They learn what should be growing on a particular site. They don't often make mistakes.

"We have to make sure that people don't work for too long under continuous sun and that they have enough water to drink," says Felicia. Rain can be dangerous because surfaces can get slippery, and there is the possibility of lightning. "We tell people working at a height to stop. Workers on rooftop projects have to be particularly careful."

For plant display work, set-up often starts at 6.00 a.m. or even 10.00 p.m. the night before; the displays have to be taken down well outside of normal working hours. "My busiest times are at the start and end of the day," says Felicia.

Felicia took a diploma in horticulture and landscape management at Ngee Ann Polytechnic. She chose this type of work because she loves plants and did not want to be desk-bound. She feels that more Singaporeans should enter the industry. "Unfortunately, many people have been brought up to see business or science or the traditional professions as the only options. Many people see horticulturists as just 'gardeners'. We have an image problem. But despite that, I love my job."

Noel Thomas is one of the NParks conservation managers responsible for Pulau Ubin, where native flora and fauna are allowed to thrive with minimum disruption. In his conservation role, Noel takes part in wildlife surveys, bird-ringing censuses, and monitoring of reptile and mammal populations. On a more mundane level, he also has to ensure that trails, roads, signboards, visitor facilities and other structures are kept in good order. There are two passions in Noel's working life: nature and sharing his enthusiasm for it. He has been fascinated by wildlife from an early age, when he would regularly volunteer at the zoo and other animal welfare organisations. He studied zoology in the United States. Upon returning to Singapore as a qualified field biologist, he found his job on Pulau Ubin to be a perfect fit.

Noel spends most days on the island, commuting by boat. After dealing with urgent emails, he checks the area on foot or by bike. "The morning hours are a good time to spot reptiles and birds. I walk the island with my GPS device and little journal, and make notes of what I see." He and his conservation officers host visits by groups of schoolchildren. "The important thing is to hold their attention. For example, when holding a beach cleanup, I begin by giving them a presentation about coastal conservation that has plenty of pictures," he explains. "It's important to take out extensive scientific facts and diagrams when working with children and deliver the message in a more exciting way." Adult visitors include tourists, locals and members of nature groups. "One thing they all have in common is that they leave Pulau Ubin pleasantly surprised."

Noel finds that children these days are better informed than they were 10 years ago, largely through the influence of the Internet. "Sometimes they have a thing or two to teach you!" he says. "Engaging the public means learning something. It goes both ways."

NParks officers assessing the health of a tree.

Left: Tree pruning (far left) and aerial inspection (top right) are normally done with the aid of a bucket lift. Sometimes when the vehicle cannot get near the tree or the lift cannot reach high enough, tree-climbing (lower right) is the only solution. Safety gear is essential.

Below left: Allowing greenery to grow on a moveable trellis permits access to the concrete structure for maintenance.

Greening the streets

From the 1970s onwards, the network of streets and roads lined by greenery steadily increased – so did the responsibilities of the people who planted and maintained them. Today, around 120 staff, plus contractors and their labourers, are backed by an annual budget of some $20 million. According to former streetscape director Simon Longman, "It's the standard of maintenance that justifies using public resources at this sort of level. It also distinguishes Singapore from some other countries that have tried to follow the greening path."

Trees need the right amounts of moisture, soil aeration and sun. They may do well on vacant land, but in the harsh conditions of Shenton Way, the story may be different. A tree that might grow to 30 metres in ideal conditions may perform less well in an urban environment.

Roots need space for aeration, so LTA provides major roads with a 2-metre planting strip on each side. On expressways and major arterial roads the centre dividers are typically four metres wide.

In some parts of Singapore, the soil condition is not ideal for large trees. Dig down, and after only 50–80 cm you may find ground water. Tree roots, which absorb oxygen and nutrients, do not like being waterlogged, and in such conditions the tree tends to form a shallow root plate. Arborists have to anticipate any possible effect on stability. As Simon put it, "Tree failures are very often soil failures in the first place."

Weather forecasters can predict severe storms up to 45 minutes ahead.

The most lethal are those known as "Sumatra squalls", which pick up energy as they travel towards Singapore across the sea. At these times, NParks staff are especially on the alert for reports of snapped branches.

The roadside trees are regularly inspected by a team of around 100 arborists, who identify problems in advance. Diagnostic equipment can detect some defects in trees. Drilling can locate decay; sonic tomographs can detect hidden cavities or poor-quality wood. A rapid-response team can get to a trouble spot within 30 minutes of getting a tree-failure report from the public. As the streetscape staff became more highly trained, incidents of snapped branches and fallen trees dropped from some 3,300 in the year 2000 to less than 1,500 annually by 2012. At the same time, the number of tree inspections was increased. But no matter how frequent the inspections are, or how expert the inspectors, it will never be possible to predict and prevent every tree failure that occurs.

Creating extra space

Trees roots need space and a good volume of soil in which to spread. This is not always available along busy roads, where drains or other utilities may be close by. CUGE researchers have tested some technically innovative solutions. One is the placing of reinforced polyethylene "cells" beneath the surface of a road (right), carpark or pedestrian walkway. These hexagonal or rectangular modules can link together to form a larger matrix, and they may even be able to support a carriageway. They are filled with an uncompacted soil mixture suitable for tree-root growth, of a kind that, without a robust surrounding structure, would not be stable enough for such a location.

Another innovation is the use of "structural soil". This is a mixture of soil and granite chips, which carries loads better than soil alone and also allows for deeper rooting, which contributes to better anchorage of trees. Although it offers less soil to the roots than structural cells, it is relatively inexpensive and the individual components are readily available.

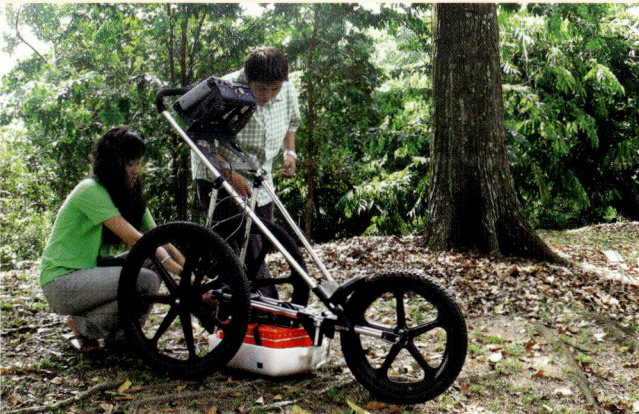

Tracing roots

No matter how closely you monitor them, root systems will occasionally cause problems. They can invade service trenches, breach water pipes or block sewers. In one instance, an Angsana root in search of water crossed a 15-metre garden and followed a sewage pipe into a toilet bowl! But such occurrences are rare. Digging up the ground to locate and trace roots is disruptive, time-consuming and expensive. CUGE has trialled the use of ground-penetrating radar (GPR) (above), which the construction industry uses for locating rebars or cracks in concrete. Compared to conventional digging, it is quick and non-invasive. Applying this technology to tree roots requires careful recalibration of the equipment, and the quality of the results depends on soil type and water content at the site of evaluation. Even though still in its infancy as an arborist's tool, GPR has proved effective in some situations.

The right tree in the right place

Not every tree is suitable for streetscape planting. Some drop all their leaves in a short period, causing inconvenience to street cleaners. Trees can present other problems. The crown of the Pong Pong (*Cerbera odollam*), for example, expands quickly and gives good shade. However, the roots tend to be invasive. More dramatically, they bear fruit the size of tennis balls (below). They can fall on passing cars and cause damage. That is why the Pong Pong is no longer planted next to roads.

● Mandai Road

● Mount Pleasant Road

● Lim Chu Kang Road

● Opposite: Various areas within housing estates are managed by NParks, HDB, LTA, NEA, PUB and the various town councils. The agencies have to coordinate with one another to ensure that greenery is maintained. Contractors have to take care not to damage infrastructure such as cables, water and gas pipes, and drains.

Heritage Roads

In August 2001, the then minister for National Development, Mah Bow Tan, announced that five roads would be designated "Heritage Roads", on account of the visual impact made by the mature trees growing along them. Arcadia Road, Lim Chu Kang Road, Mandai Road, Mount Pleasant Road and South Buona Vista Road were officially gazetted in 2006. Heritage Roads have a green buffer of up to 10 metres on each side, within which removal of trees or plants is not allowed. Heritage Road status has to be taken into account in planning development works, and even trees along non-Heritage Roads are treated with consideration.

A living heritage

In 1991, two large areas (Central Singapore and Changi) were designated Tree Conservation Areas (TCAs). In that first year alone, over 800 mature trees threatened by development proposals were saved.

The year 2002 saw the launch of the Heritage Tree scheme. The initial candidates were nominated by NParks staff. Subsequently, nomination was opened up to the public. By the end of 2012, there were close to 200 trees on the Heritage Tree register. Lightning strikes occur in Singapore on 270 days a year on average, and therefore, owners of tall and mature trees are advised to install lightning conductors on their trees. NParks showed the way by putting them on their own trees first. One big Tembusu in Singapore Botanic Gardens registered three lightning strikes in eight years.

In the case of Heritage Trees – as with all conservation issues – different people have different priorities, as LTA found. Chief executive Chew Hock Yong explained, "For example, in cases where a Heritage Tree might obstruct a double-decker bus, LTA has worked with bus operators to run single-deckers instead or re-route the service. This is not always popular. Different stakeholders have their own priorities, be they MPs, grassroots representatives or residents asking why a place is not being serviced. There are usually many sides to the argument. Such decisions involve trade-offs."

● Rainbow Gum (*Eucalyptus deglupta*), a Heritage Tree in Katong Park.

Watching over the trees

Since 1997, NParks has used a geographic information system to monitor the health of the tree population under its care. NParks officers record and update the species, size, location and condition of every tree during their site inspections. Managers can use the data to keep track of trends, such as the increasing diversity of tree species.

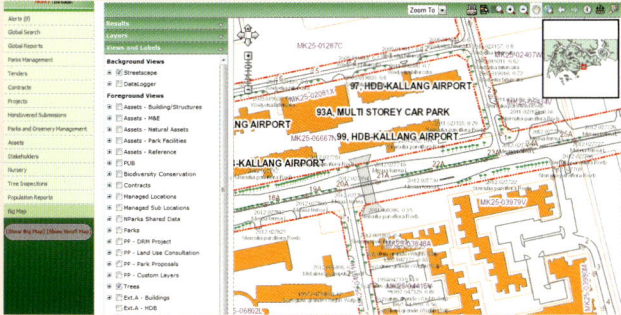

"It is our roadside greenery and how we have maintained it over the decades that set us apart. All cities have parks; some bigger than ours. But few have roadside greenery that is as lush and extensive as ours."

Poon Hong Yuen, chief executive officer, NParks

More trees, more variety

In 1997, when NParks first started using the geographic information system, it revealed that 75 percent of the tree population in the parks and along roadsides was made up of trees from only 20 species (below left). By 2013, that number had doubled to 41 (below right). The total number of tree species has grown to over 700.

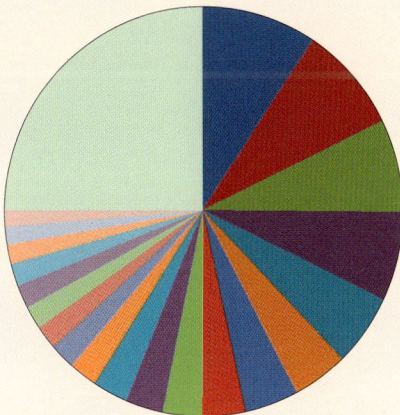

1997

- Samanea saman
- Peltophorum pterocarpum
- Swietenia macrophylla
- Syzygium grande
- Archontophoenix alexandrae
- Khaya senegalensis
- Cinnamomum iners
- Tabebuia rosea
- Adonidia merrillii
- Arfeuillea arborescens
- Mimusops elengi
- Roystonea oleracea
- Cassia fistula
- Syzygium polyanthum
- Pterocarpus indicus
- Khaya grandifoliola
- Xanthostemon chrysanthus
- Mangifera indica
- Terminalia catappa
- Roystonea regia
- Others: approx. 400 species

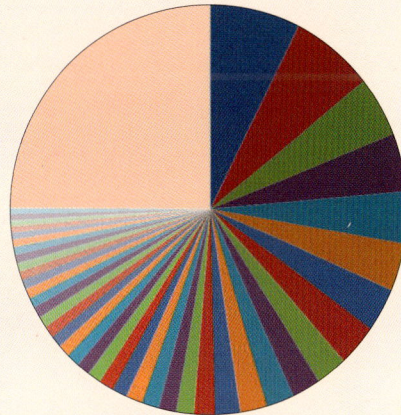

2013

- Samanea saman
- Peltophorum pterocarpum
- Swietenia macrophylla
- Syzygium grande
- Khaya senegalensis
- Archontophoenix alexandrae
- Tabebuia rosea
- Hopea odorata
- Mimusops elengi
- Cinnamomum iners
- Arfeuillea arborescens
- Adonidia merrillii
- Roystonea oleracea
- Syzygium polyanthum
- Syzygium myrtifolium
- Cassia fistula
- Khaya grandifoliola
- Pterocarpus indicus
- Fagraea fragrans
- Mangifera indica
- Planchonella obovata
- Roystonea regia
- Terminalia catappa
- Cocos nucifera
- Callistemon citrinus
- Cratoxylum formosum
- Lagerstroemia speciosa
- Sandoricum koetjape
- Erythrophleum guineense
- Livistona chinensis
- Alstonia angustiloba
- Syzygium lineatum
- Dalbergia latifolia
- Callerya atropurpurea
- Filicium decipiens
- Plumeria rubra
- Melaleuca bracteata
- Maniltoa browneoides
- Kopsia flavida
- Casuarina equisetifolia
- Others: approx. 700 species

③
WHO ARE THE PARKS FOR?

Parks exist to make life better for Singaporeans. Wherever they live, people need easy access to green spaces where they can pursue a whole raft of leisure interests, from birdwatching to mountain-biking to kite-flying to jogging. In response, the emphasis of park design has evolved, putting people firmly at the top of a planner's priority list.

● A kite festival at West Coast Park.

Attracting people

In the design and management of parks in the 1970s, ease of maintenance came first. With the building of New Towns in the 1980s, and even more so after NParks was formed in the 1990s, that changed. Maintenance mattered, but park designers became more creative. This showed not only in the town parks but also in major redevelopments at Marina City Park (where Gardens by the Bay is today), Pasir Ris Park, Fort Canning, Mount Faber, Singapore Botanic Gardens and elsewhere.

The emphasis on variety extended to the roadsides – for example, it was the thrust of the Streetscape Greenery Master Plan, drawn up in 2002, in consultation with the Singapore Institute of Landscape Architects (SILA), to create unique identities for clusters of roads at strategic locations.

In Singapore's early days, there was a shortage of creative expertise. As the 1980s dawned, the then prime minister, Lee Kuan Yew, brought in some Japanese landscape architects. One of them was Jun-Ichi Inada, who joined the PRD in 1983 and became director of planning and development for NParks in 1991. He headed many successful projects in Singapore, including development of the Singapore Botanic Gardens, Bukit Timah Nature Reserve and Pasir Ris Park. He also designed the National Orchid Garden and Jacob Ballas Children's Garden.

Today, local parks are designed with as much care as the bigger destination parks – they are close to where people live and account for half of all park visits. Each one is designed individually, taking into account the local demographics.

A park in newly built Punggol will be different from one in Tiong Bahru, which has a higher proportion of elderly residents.

Both NParks and HDB regularly survey the way parks and gardens are used. More and more people are using fitness corners and jogging tracks – those using them at least once a week doubled between 1998 and 2008. Putting fitness corners close to children's play areas also encourages the generations to mix.

It is not easy to design a park for many different tastes and uses. As Kong Yit San, assistant chief executive officer, Parks Management and Lifestyle Cluster, NParks, said, "People want every kind of facility close to their own home. All park users have to show consideration for the interests of others."

The private sector

The year 1993 saw the formation of a Landscape Design Panel, bringing together creative expertise from the private and public sectors. Today, the private sector plays an increasingly important role. As with nurseries and maintenance contractors, since the 1990s, much of the landscaping design for parks has been contracted out to commercial firms. An important force in their industry is the Singapore Institute of Landscape Architects, which has also been successful in promoting Singapore's landscaping services overseas.

The modern adventure playgrounds at Pasir Ris Park (opposite, below) and Woodlands Waterfront Park (above) are more sophisticated than the row of swings at Kallang Park (opposite, above) in 1965.

Art in three dimensions

The design of a park can be lifted to another level by works of art or decorative pieces, such as sculptures. In the 1980s, Singapore's former chief minister David Marshall donated three sculptures by Sydney Harpley to the Singapore Botanic Gardens: Girl on a Swing (1984), Girl on a Bicycle (1987) (left, top) and Lady on a Hammock (1989), all dedicated to the children of Singapore.

In 1988, the government introduced tax incentives to encourage private donors to support public sculpture. Three years later, the then prime minister Goh Chok Tong unveiled a group of sculptures at Marina City Park. Donated by a food manufacturer, these eight legendary Chinese heroes represent loyalty to country, filial piety, benevolence, love, courtesy, righteousness, honesty and modesty. When the park made way for the development of Gardens by the Bay, the group was moved to the Chinese Garden.

Other notable works include a sculpture of a mother and baby by Dr Ng Eng Teng at Tampines Central Park and sculptures of a Chinese lantern procession at Telok Ayer Green (left, below).

In Sengkang New Town, art lovers can enjoy Sengkang Sculpture Park, a collection of works with a marine theme, designed so that the public can interact with them.

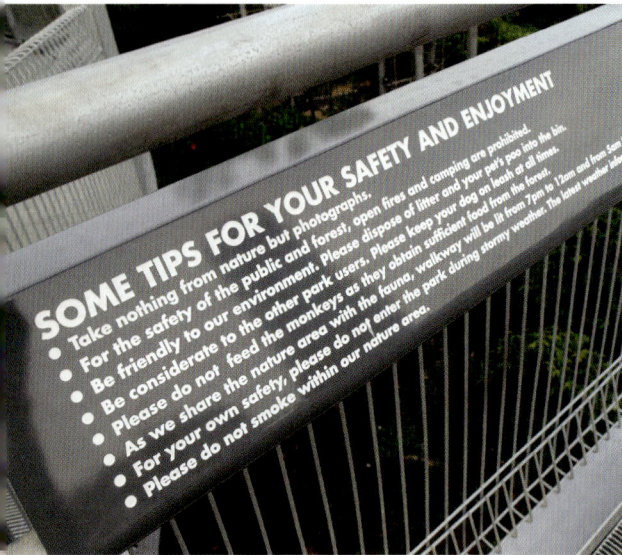

SOME TIPS FOR YOUR SAFETY AND ENJOYMENT
- Take nothing from nature but photographs.
- For the safety of the public and forest, open fires and camping are prohibited.
- Be friendly to our environment. Please dispose of litter and your pet's poo into the bin.
- Be considerate to the other park users. Please keep your dog on leash at all times.
- Please do not feed the monkeys as they obtain sufficient food from the forest.
- As we share the nature area with the fauna, walkway will be lit from 7pm to 12am and from 5am to 7am.
- For your own safety, please do not enter the park during stormy weather. The latest weather information.
- Please do not smoke within our nature area.

The art of communication

There was a time when NParks signboards consisted of a long list of rules, regulations and prohibitions. Many older parks staff agree that these contained too much information to be effective – in fact they were off-putting.

So, how much information should these boards contain?

As people have become more educated in the use of parks, a more subtle approach has been adopted, using a clear system of graphics.

Today, a red circle with a diagonal red slash is a prohibition. A red triangle warns of a danger ahead. A red border identifies advice, while a sign without a border provides educational information. But there are variations to how an educational sign may appear. Information and education can be more effective than prohibition signs, for example, in discouraging people from feeding animals.

A fruity theme

Some early park signage was very creative. In Yishun, for example, it was designed to look like fruit. The signs were huge, almost like sculptures, if not exactly artistic. It was difficult to give the "rambutan" realistic "hairs", so it was renamed "pulasan" – similar in appearance but a little larger and without the spines! Making such objects requires craftsmen, who are in short supply. Parks managers had to replace these "sculptures" with conventional signs – less fun but more practical.

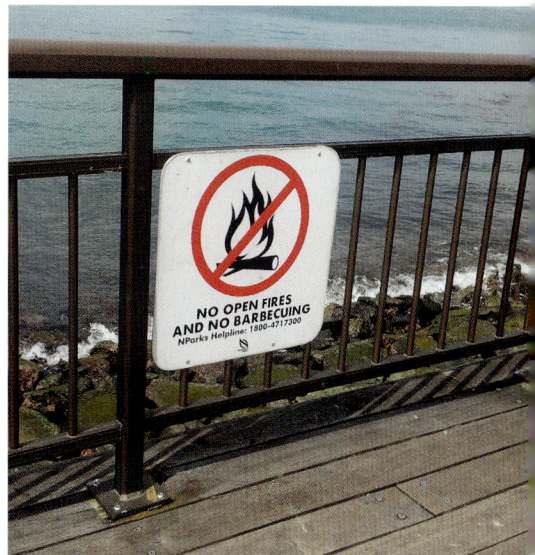

● Opposite: An etiquette sign dishes out advice and tips to visitors (left) and a colourful map board guides visitors through the sights of a park (right).

● Above: An educational sign explains the what, why and how of dealing with monkeys (left); advisories inform visitors what to do or to look out for when entering the park (centre); and a prohibition sign warns visitors of what not to do in a park (right).

● Below: Interpretative signs such as this one at Singapore Quarry provide information on the history and significance of an area.

Tiong Bahru Park

The Singapore Improvement Trust, forerunner of the Housing and Development Board (HDB), constructed a housing estate at Tiong Bahru before World War II. Several blocks of old flats from the period, built in Art Deco style, have been granted heritage status and are sought after by discerning buyers today. Construction continued after the war.

However, by the early 1950s, the population had grown, outstripping the available housing. A series of devastating fires, the last in 1961 at Bukit Ho Swee, broke out among overcrowded kampong houses, making tens of thousands homeless. This gave a boost to the building programme of the newly formed HDB, which immediately set about constructing flats at Tiong Bahru for lower income groups.

Against this background, Tiong Bahru Park was developed in 1967. It has been upgraded since, most notably in 2000, when a "tilting train" arrived at the adventure playground.

● Above: The fire at Bukit Ho Swee in 1961 jump-started residential development at Tiong Bahru.

● Below: Palms and aquatic plants at Tiong Bahru Park.

Children's favourites

Older Singaporeans think back nostalgically to the playgrounds designed by HDB in the late 1970s to 1990s, featuring "sculptures" in the stylised shape of a dragon, pelican, tortoise, rabbit, elephant or dove. Fruit – watermelon, mangosteen and pineapple – was another theme. The early history of Singapore was the inspiration for a rickshaw, trishaw and sampan. More recently, playgrounds have featured dinosaurs, a crocodile (left, formerly at Tampines Sun Plaza Park), a kangaroo and a teapot with mushrooms. A modern favourite is the tilting train at Tiong Bahru Park (below).

Sadly such features do not always measure up to modern safety needs: steel can go rusty; worn mosaic tiles develop sharp edges. Today's parents are not willing to see their children exposed to risks that were routine a couple of generations ago. So, most modern playgrounds are fitted out in plastic rather than metal or other hard-edged materials. NParks and HDB install equipment certified for safety according to international standards.

Toa Payoh Town Park

It is hard to imagine today, but Toa Payoh was once an area of kampongs, squatter settlements, and fish and vegetable farms. The name comes from a combination of Hokkien and Malay and means "big swamp". In the early 1960s, the government encouraged the people who lived there to move into housing estates. The development of Toa Payoh began in 1964 – it was the first new town to be built from scratch by HDB and featured a community plaza, a library block and a town garden.

The designers gave Toa Payoh Town Garden (as it was then known) an "Oriental" theme, by creating a lake with a series of artificial islands linked by bridges. It soon became a favourite setting for wedding photos.

The name "Toa Payoh Town Park" dates from 2002, when the park re-opened after partial closure to make way for a temporary bus interchange during construction of the HDB Hub. The trees around the edge of the park show how effectively greenery can screen a park from busy roads, creating a pleasant, relaxing environment. The park gives easy access to one end of the Whampoa Park Connector.

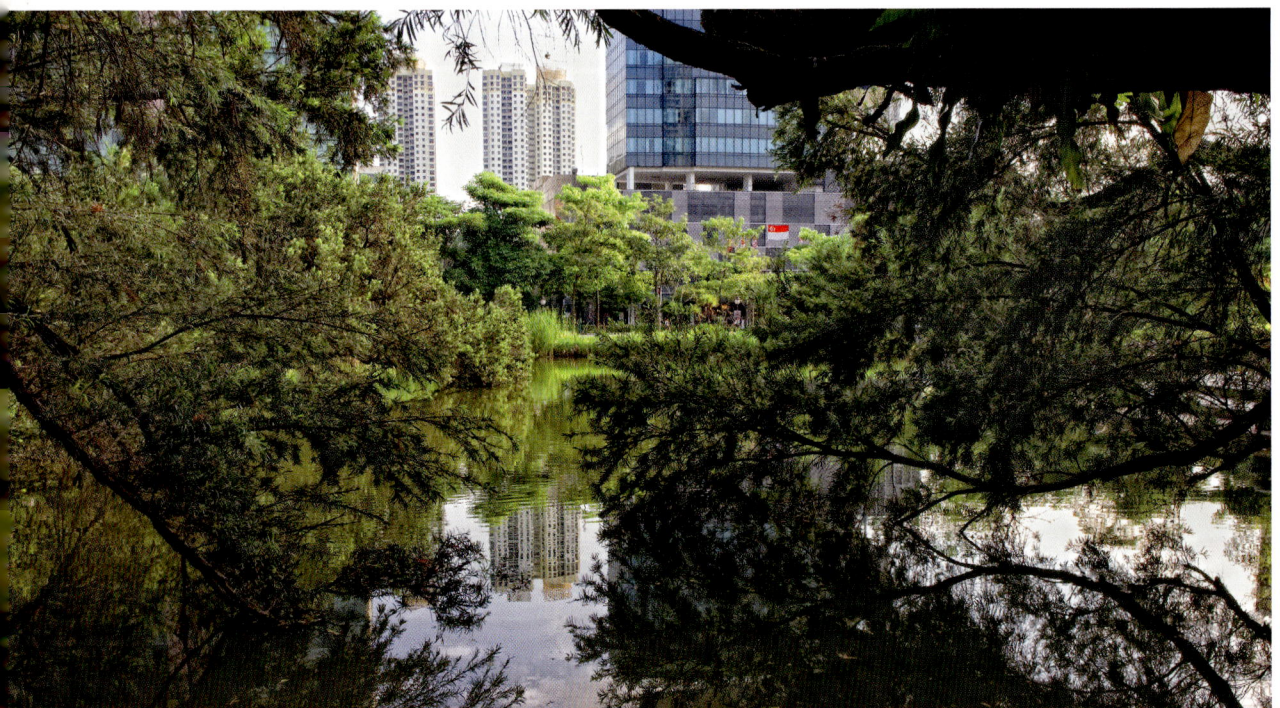

Dragonflies, masters of aerial manoeuvre

Ponds and waterbodies are diverse habitats and therefore host a range of dragonflies and damselflies. In all, 127 dragonfly and damselfly species have been recorded in Singapore.

Dragonflies spend the greater part of their lives in water as larvae, or "nymphs", before emerging as adults. The best time to watch them is during sunny weather from mid-morning onwards, when they will be active after warming up in the sun. Different species will visit ponds at different times – it is worth waiting to see which ones appear. Their flying abilities are amazing – they can fly upwards, downwards, forwards, backwards and sideways. In fact, dragonflies were the original inspiration for helicopters.

The pond at Toa Payoh Town Park shows how nature can prosper when surrounded by urban development. It is the perfect habitat for larvae, and 18 species have been recorded there. Two are particularly numerous: the Common Scarlet (*Crocothemis servilia*) one of the largest red dragonflies (right above), and the Common Parasol (*Neurothemis fluctuans*). The sharp-eyed visitor may spot the Variable Wisp (*Agriocnemis femina*), a tiny damselfly, or another damselfy, the Ornate Coraltail (*Ceriagrion cerinorubellum*) (right below), a voracious predator of other insects. A rarer species – at least in Singapore – is the Orange-faced Sprite (*Pseudagrion rubriceps*).

● Opposite, top left: The iconic observation tower at Toa Payoh Town Park.

● Opposite, below: The lake at Toa Payoh Town Park is surrounded by Bottlebrush Trees.

● The bridge, with its repeated hexagon motif (opposite, top right), and the ponds linked by picturesque bridges (below) express an "Oriental" theme in the landscaping.

Ang Mo Kio Town Gardens

Plans for a new town at Ang Mo Kio, drawn up in the early 1970s, proposed the creation of two town parks, each on a raised area of ground: Ang Mo Kio Town Garden East and Ang Mo Kio Town Garden West.

When Ang Mo Kio Town Garden West was completed in 1982, it was the largest HDB town garden at the time. The designer retained much of the natural vegetation, enhanced by a lotus pond, a pergola and an observation deck. The natural slopes give the garden much of its character. It is further enriched by the wildlife that flourishes there, including butterflies and dragonflies.

Ang Mo Kio Town Garden East was built on a former rubber estate. It still features not only rubber trees but also other economically important species, such as cinnamon and nutmeg.

● Royal Palms (*Roystonea regia*) in Ang Mo Kio Town Garden West.

● Right, above: The Painted Jezebel (*Delias hyparete metarete*), a butterfly often seen up in the canopy of tall trees.

● Right, below: Sculpture in the form of a rubber tree seed, recalling the rubber plantation formerly on what is now Ang Mo Kio Town Garden East.

Bishan-Ang Mo Kio Park

For the flagship project of PUB's ABC Waters programme, Bishan-Ang Mo Kio Park, NParks and PUB worked together to transform an upper reach of the Kallang River from a concrete canal into a meandering, natural-looking stream. A combination of plants, natural materials and civil engineering techniques were introduced to soften the edges of the waterway, give it a natural appearance and prevent soil erosion. This created a habitat for a variety of aquatic and bird life. Dragonflies and damselflies are present in great numbers, representing over 20 species.

During dry weather, the flow of water is confined to a narrow stream in the middle of the river. In the event of a storm, the adjacent park area doubles up as a conveyance channel, carrying the rainwater gradually downstream.

Visually, the scale of the park, with its backdrop of high-rise housing blocks, is dramatic. In size, the park ranks not far behind Singapore Botanic Gardens – including the areas on both sides of Marymount Road, joggers can run a lap of over 5 km.

● Above: The park offers many recreational possibilities, in particular, dining. The many options draw crowds, especially on weekends.

● Below: The canal before it was turned into a naturalised channel.

● Bottom: Nature recreated in urban surroundings: Bishan-Ang Mo Kio Park.

● Opposite: The sharp-eyed, both avian and human, will find an abundance of aquatic life living in the meandering stream.

"It's refreshing to have meals at a cosy restaurant set in a park environment – green, clean, relaxing. I bring my kids to cycle, roller-blade and take a brisk walk round the park, before eating at the restaurant."

Lee Leng Leng, mother

Destinations in their own right

As the public transport network and number of cars grew, people became more mobile. In their leisure activities, they wanted choice. These were factors behind plans to develop three big parks into "destination parks", announced in 2012. Located in the east, west and north, each will have a theme and attract visitors not only from the nearby neighbourhood but also from all over Singapore. The public were asked for their ideas – more playgrounds, big slides for children, and water features were on the list.

Jurong Lake Park, in the west of Singapore, is home to a third of Singapore's resident bird species. Water is the theme. One proposal is to turn grassy areas beside the lake into islands filled with play equipment, linked by bridges. The lake, with its boardwalks, fountain and fishing points, will be a magnet for strollers, joggers and cyclists.

● Above and left: The tranquil waterside environment of Jurong Lake Park.

Admiralty Park, the biggest park in northern Singapore, contains the largest nature area in any of the country's parks other than the nature reserves. Running through it is a river, Sungei Cina, and a boardwalk gives a good view of the mangroves and a freshwater swamp. The hilly terrain gives scope for creatively designed features such as big slides and places for children to climb.

● Top: Naturally regenerating forest at Admiralty Park.

● Centre left: A Giant Mudskipper (*Periophthalmodon schlosseri*) at Sungei Cina.

● Centre right: Fruit of a Nipah Palm (*Nypa fruticans*), a lover of waterlogged ground.

● Left: From the West Entrance, a footpath leads visitors to the nature area ahead.

East Coast Park

East Coast Park is the biggest of all the destination parks. In fact, it is the largest park in Singapore, attracting over seven million visits a year.

It was created when HDB reclaimed a long strip of land along the coast from Bedok to Tanjong Rhu. This massive project, which started in 1966, involved moving tens of millions of cubic metres of earth to the seashore on conveyor belts. The result was the opportunity to build an entirely new community at Marine Parade. The first flats were finished in 1973. The HDB handed over East Coast Park to the PRD, forerunner of NParks, in 1976. Soon, the park boasted trees, paths and waterbodies. It became Singapore's favourite spot for outdoor relaxation and exercise, with facilities for a wide range of leisure activities such as cycling, jogging and, nearby, sailing.

● Opposite: A tranquil spot shaded by Coconut Palms (*Cocos nucifera*), with an ever-changing panorama of container ships.

● Top: The then prime minister, Lee Kuan Yew, inspecting the plans for East Coast Park (left), and the progress of reclamation works (right), 1972.

● Above: There was no mature vegetation on the newly reclaimed East Coast Park in the 1970s (left), a contrast to the lush greenery cladding the park and the highway today (right).

● Below: Cycling, in-line skating, barbecuing and water-skiing are just some of the activities that make East Coast Park the most popular park in Singapore.

Opening up a new dimension

The idea of hundreds of kilometres of green corridors for walking and cycling would have seemed incredible 30 years ago. However, in 1991, the government approved a proposal for such a network of greenways – now known as the Park Connector Network or PCN. The total length of the network was to be over 300 km and completed over 20 to 30 years. The first stretch was opened in 1995: the Kallang Park Connector, linking Bishan-Ang Mo Kio Park to Kallang Riverside Park.

Planning for the network in a densely built up city-state is no easy feat. NParks planners have to contend with constraints of existing development sites, while keeping in mind the main objective of linking major parks and nature areas. In order to optimise land use, the network has been designed to run on covered drains, along drainage reserves, under MRT viaducts and even into adjoining public housing land. In an urban setting, the network is inevitably interrupted by roads and bridges. NParks is working with LTA and other stakeholders to improve connectivity. In an effort to create a sense of spaciousness in a highly urbanised landscape, the planners located the network along the coastline or waterbodies where possible.

The network is gaining in popularity among the public. A 2012 NParks survey showed that visits to the network increased substantially from 4 percent in 2007 to 27 percent in 2008, the year the first park connector loop was completed. While the main aim of the network is to link parks together and provide recreational spaces for all, residents have come to appreciate the network as a more scenic route to markets, schools and even transportation nodes.

The network has also helped maintain Singapore's biodiversity. By linking existing green areas, it has countered the habitat fragmentation that threatens wildlife populations. Special thought is put into planting the right type of trees and shrubs along the network, as birds and other wildlife look for food in the form of nectar, fruits and insects. The results speak for themselves – a recent survey logged 90 bird species, 57 butterfly species and 22 dragonfly species. When the vegetation fully matures and becomes more complex, the effect on wildlife will be fully revealed.

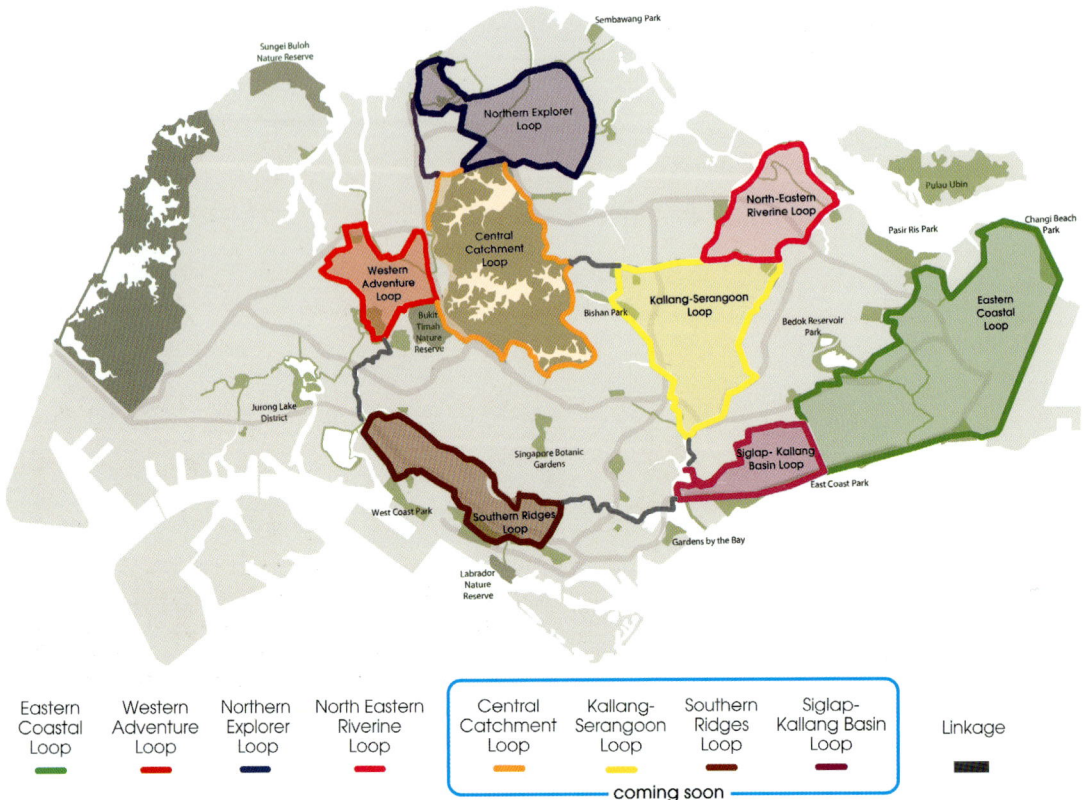

"Safe and smooth – park connectors really give us cyclists a route that is meant for us. This way we can see and feel the beautiful surroundings in safe conditions."

Nick Phoon, engineer

● Opposite below: The Park Connector Network today.

● Above: Joggers, cyclists and in-line skaters enjoy the Alexandra Canal Linear Park, which was built on top of a covered canal.

● Below: Excited participants at the launch of the Western Adventure Loop, 2009. The loop links eight parks in western Singapore.

●●●●● Round Island Route

Planned for the future is a continuous walking and cycling route encircling the island, taking in many parks and places of historic or cultural interest, and passing close to the homes of over half the population. One complete circuit of the Round Island Route will be over 150 km long.

Activities under the sun

Modern life may sometimes seem stressful, but the rich variety of nature reserves, parks and gardens in Singapore gives everyone an opportunity to take a break. There is a surprising amount of space for outdoor activities in Singapore. The range of opportunities is enormous. A visit to the NParks website (nparks.gov.sg) and Facebook (facebook.com/nparksbuzz) is a good way to find out what to do and where.

"If you want to see birds, it's best to go in a small group, because that gives you more ears and eyes. A big group is too noisy. If you want to see ground birds, it's best to go on your own, or with just one or two others – they are very shy."

Alan OwYong, past chairman of the NSS Bird Group

"Looking at nature gives you an appreciation of a slower pace of life. It's easy to go into the virtual world, making friends online. It's more difficult to slow down your pace of life and actually have an awareness of your surroundings. But if you take a walk, and examine the birds and the insects all around, you can really connect. This is NOW."

Chua Siew Chin, NParks volunteer

"Forest Adventure@Bedok was very exciting! When I'm tall and brave enough, I want to try the Grand Course too."

Erin Chen Si Ying,
primary school student

"An activity that I enjoy is sitting by the café at Punggol Park having a cuppa with my family and friends. The tinkling of the cutlery and utensils accompanied by laughter from diners and children running around bring life and add a different facet to the park. A park in the heart of a housing estate brings delight to one's spirit."

Jacqueline Tan,
avid park visitor

"I was a birdwatcher from a young age. What really compelled me to try mountain biking was that it is another way of exploring the forests. Biking is a workout. You feel really good. Also, as you move through the forest on one of the dedicated trails, it's quiet. You don't really talk. Even if you're cycling as a group you focus on the trail, watching your line, feeling the feedback from the ground. One of the special things about Singapore is that you don't need much planning to go mountain biking. The trails are very accessible."

Debby Ng, photojournalist

"I practise yoga at East Coast Park. The park becomes my teacher. It is a different way of liberating cognition, experiencing blissful mental states when I accept the busy Singapore Strait as an integral part of sun, sea, sky and beach on a simple sheet of yoga mat."

Kenryuu Ong,
regular park visitor

"Going to Sungei Buloh sparked off my interest in bird photography in the late 1980s. I was surprised to see huge numbers of migratory birds flying around in front of me. I couldn't imagine that in this tiny island we could see such a wonderful sight – I thought it could only happen in places such as the US, Australia or Africa. That's what inspired me to take up bird photography. One thing I discovered is that you do need good equipment, but it was worth the investment."

Lee Tiah Khee,
professional photographer

"During year one at polytechnic, my friend pulled me into kayaking. Then I got really interested because it's fun. You could say that I got in for social reasons, then after that it turned into a passion. When you're on the boat, you're moving; the boat seems to move very fast because it's cutting through water. You feel very nice. The scenery at MacRitchie is beautiful, particularly at sunrise and sunset. You have the sense of an escape – you feel more at ease. It's a place where you can go and get away from everything."

Muhammad Radi,
full-time national serviceman

"There is no better feeling than jogging in the serene setting of the park. It's not just the peace and tranquillity it offers but also the aroma of the flora and fauna that relaxes me, frees my mind and energises me for the day's challenges."

Mano Mahendran, director

"We come here (Punggol Point Park) for sand play and cycling, to be close to nature and enjoy the birds singing. My son's a toddler – he gets to play with other children, and it is good for his social skills. He also gets to learn about the world around him from direct experience with nature."

Boon Wei Lek, mother

"I like to go to Pasir Ris Park for picnicking, cycling or just taking a walk. Not only is it near my house, but it is an opportunity for me to reflect or share thoughts with my friend. Also, the natural wind and the sound of the waves calms me down and I love that feeling."

M. Bhairavi, student

"I reward myself for getting up early on Sunday morning by skating along the beach at East Coast Park. From the lagoon to Changi, the skating track affords a relaxing view that takes your breath away with varying levels of incline to keep you on your feet."

Damon Wong, skating enthusiast

Keeping the garden clean

A study of littering by the National Environment Agency (NEA) showed that many factors, including the type and location of bins, play a part in shaping behaviour. For example, mothers are more litter-conscious than fathers, particularly when setting an example for children. Some young people, possibly because of peer pressure, seem indifferent to the effects of littering, even excusing it on the basis that it gives employment to low-paid workers.

The amount of littering varies according to the way a park is used. Visitors to the Singapore Botanic Gardens and other iconic parks tend not to litter; leisure activities at East Coast Park generate more mess, as do concerts and other performances. In 2010, nearly 10,000 littering fines were imposed by NParks – an average of nearly 200 a week. NEA's CEO, Andrew Tan, sees value in the "broken window theory" – that once the environment starts to degrade, the process will accelerate. "We want to start the day with a clean environment and sustain that for the rest of the day."

When developing nature parks and reserves, NParks decided not to install litterbins, in keeping with the natural surroundings. The public was trusted to keep their side of the bargain and take out of the park whatever they brought in.

4

ANYONE CAN BE A GARDENER

The gardening experience is open to anyone – even the flat-dweller. Some people grow plants indoors, and others join community gardening groups. The important thing is to get actively involved. Some people want to recapture some of the simple pleasures of kampong life; others find an almost spiritual satisfaction in getting close to nature. And no one benefits more than children – gardening is a powerful educational experience.

● You are never too young to learn about gardening.

What, gardening in Singapore?

The public has been involved in the greening of their surroundings ever since the first efforts to turn Singapore into a garden city. There was Tree Planting Day, which started formally in 1971. The first Horticulture and Aquarium Fish Show took place in the same year at the carpark of the Ministry of National Development and continued at intervals at various larger venues until the mid-1980s. More recently, the Singapore Garden Festival attracted gardeners in their thousands.

Hands-on activities are part of many educational and volunteer programmes. The Community in Bloom movement is in that tradition. It started in 2005, at Mayfair Park Estate. Soon, gardens were created in housing estates, schools, hospitals, welfare homes, places of worship, offices and factories. Within two years there were 200 groups; by 2012 there were over 500. With advice from NParks, the green-fingered from Woodlands to Bukit Merah, from Jurong West to Pasir Ris, were communicating not only face-to-face but via blogs and other social media platforms. They were gaining first-hand knowledge of plant identification, propagation, watering, pruning, weeding, soil types and the use of fertilisers.

● Opposite, top left:
Community gardening at
Woodlands Zone 2 RC.

● Opposite, top right: At
Coral Ris RC, a resident
teaches a student from a
nearby school how to garden.

● Inset: The Sponge Gourd
(*Luffa aegyptiaca*) is a
popular vegetable.

● Right: The garden at Coral
Ris RC.

The Gardeners' Cup

Community gardening was one
of the themes of the Singapore
Garden Festival. For the 2012 show,
30 community garden groups
formed themselves into teams to
compete for the Gardeners' Cup.
Five gardens were on display. The
overall winner was "The House by
the Mangroves" (left). The gardeners,
led by Zuhir Taib of the Tampines
Starlight Residents' Committee,
recreated an attap house, calling
to mind Singapore before the time
of Raffles. Accessories included a
cooking stove and fishing nets, and
a wide variety of mangroves and
other plants.

Community in Bloom Ambassadors

Community in Bloom Ambassadors are volunteers nominated for their enthusiasm and willingness to go the extra mile in helping others enjoy the world of plants. Each year, a new group of CIB Ambassadors is nominated by other CIB Friends or by members of the general public. They devote their energy to engaging the public through road shows, exhibitions, community events, websites and blogs. Many of them have become real experts, providing gardening tips through articles in the media or on the web. They love sharing their passion, acting as willing hosts for visitors to their community gardens.

Through their efforts, more and more people are discovering the joy of hands-on gardening.

Ng Jia Wei became the youngest CIB Ambassador when he received the award in 2011, at the age of 13. A keen photographer, he helps run the garden of Woodgrove Zone 2 Residents' Committee, where the focus is mainly on fruit and vegetables, including corn, grapes and watermelon. The residents are also experimenting with hydroponics.

The oldest group member is around 86, strong and healthy. Most of them are much younger. According to Jia Wei, "I joined when I was in Primary 6. Since then I have learnt a lot from the older gardeners. Community gardening for younger people is very good. Gardening is not just about planting. We get to learn more about the community. It inculcates strong life values in us. We are developing patience, watching the plants grow with love and care." He takes plants to school, where his schoolmates take turns to water them.

Kamisah Bte Atan leads the gardening club of Jurong Central Zone D Residents' Committee. The "hardscape" of their garden, at Jurong East Avenue 1, was provided by HDB. "The soil was clay, and it needed a lot of tilling, manure and compost." Kamisah says.

They have over 30 members. About ten are particularly active. "Most of them are local residents. People just pass by and see us gardening, so we ask them to join us. They like the community feeling." Members discuss what plants to grow – it might be long bean one season, then sweet potato, then tapioca – vegetables are the most popular.

Gardening keeps the members fit and active. Many of them are quite elderly and feel nostalgic for the kampong days. One woman used to be a farmer. They enjoy showing plants to the younger generation, when school groups visit. "We get some help from the younger volunteers," Kamisah says. "We have to teach them to recognise weeds, and they get a bit scared when they first see an earthworm. But they learn. Some of the members bring their grandchildren, and they work together."

The same group runs HDB's first wheelchair-friendly sky garden, on the roof of a multistorey carpark nearby. The beds are raised about a metre above the ground, so they can easily be reached by children, the elderly and people in wheelchairs or on stools. Regular watering is essential – due to the breezy conditions, soil can dry out easily. However, despite the challenges, sunflowers and several kinds of vegetables are doing well.

Ismail bin Haji Suratman is chairman of Tampines Starlight Residents' Committee. Their "Starlight Harmony" garden is at Street 71. When Ismail got involved, the gardening group needed re-energising. He thus focused on competing for CIB Awards. In 2012, they were part of a combined team that won the Gardeners' Cup at the Singapore Garden Festival with their display "House by the Mangroves".

As Ismail explains, "It's not just about gardening itself. I just wanted to bring the members to the next level. I just want to expose them to knowledge, friendship and anything about plants."

The emphasis in the Starlight Harmony garden is on herbs and medicinal plants. "We chose them because we want to show the younger generation the types of plants that were grown in the past, in kampong days, by all the communities – Chinese, Indian or Malay. You hardly see them now. This is something not many other gardens have." The garden has its own Facebook page, and shares knowledge with friends all round the world.

Ismail applauds efforts to involve schoolchildren. "Community in Bloom gets primary and secondary schools involved and even higher institutions.

I would like to see more preschools invited to take part in competitions. The reason is not about the winning. The reason is to nurture them, to stimulate their love of nature. Life in Singapore is fast-paced. Gardening teaches us to be more patient."

Philip Li is programme coordinator with the Association for Persons with Special Needs (APSN) Centre for Adults, an employment development centre for young adults with mild educational difficulties. He went on crash courses with plant nurseries to acquire knowledge of gardening to pass on to his students.

Through gardening, Philip aims to give young people enough experience and self-confidence to earn their own living as independent members of society. His students number between 20 and 30 at any one time. He finds that many nurseries, landscape contractors and farms (yes, there are still over 200 farms in Singapore) are willing to employ people with a mild intellectual disability so long as they have some basic skills.

The garden, which includes five hydroponic units, grows mostly fruits and vegetables. Tomatoes, strawberries, apples and dragonfruit are favourites. "Many of our plants were donated," Philip explains. "When we started out we asked people to let us have plants they didn't want, however sickly. We became known locally as a 'plant hospital'."

One of his former students is now employed as a teaching assistant, managing visual aids under the guidance of a teacher. "Gardening has changed his life," Philip says. Other students just enjoy guiding visitors around the premises. One of the happy by-products of the garden is the involvement of local residents, including school students. One neighbour was so inspired that he installed his own hydroponic system.

Taking a gardening break

In 2008, Singapore Technologies Kinetics Ltd, together with the rest of ST Engineering Group, began a programme to plant 2,008 trees in Admiralty Park. Yeap Khek Teong was on the organising committee. Then, he explained, "Top management decided to make small plots of land available to staff for community gardening. Five plots were designated in five different company locations." The gardening instinct was already there. "When we started, we noticed pockets of individual planting on company land, including a papaya tree not planted by the organisation. It was a spontaneous thing, not organised. We decided to provide an opportunity for the gardeners."

NParks helped with the initial design and the building of the gardens and with the initial plant choice. "Management made some money available for plants, fertiliser, tools, etc; our members provided the labour." The first garden opened in 2009.

Having gardens near the workplace, close to common areas such as the canteen, is a good arrangement. "It's sustainable," explains Khek Teong. "People can do a bit of gardening at times when they feel most energetic. They can learn about the plants, do a bit of weeding, plant a few vegetables. And those who are not interested in participating enjoy the gardens as a place to relax."

● Top: Staff building their first community garden at the Jalan Boon Lay premises in 2009.

● Above: A green wall. Recycled elements include white safety helmets.

● Opposite, top: Eighty staff rolled up their sleeves and landscaped the front of their work premises in 2012.

● Opposite, below: The community garden, called "Garden in Bloom", incorporates wall bricks from a nearby demolished building. Vertical planting using pots is an ingenious way of saving space.

HortPark

The role of HortPark, which opened in 2008, is to provide ideas and information for the active gardener, rather than the spectator. It caters to members of the public who have gardens, large or small, on balconies, indoors or at ground level. It features displays of plants, garden layouts and garden-related products, aimed at both amateurs and professionals working in the landscape industry.

The display gardens illustrate many different themes. Preschool children and young science students can enrich their knowledge of plants and the ways they disperse their seeds, as well as propagation methods, at the Pitter Patter Potter Garden. Balinese and water gardens include do-it-yourself aquatic plant features. A herb garden introduces visitors to medicinal and culinary herbs and spices, some of them traditionally associated with mystical powers. There are planting suggestions for rooftop gardens, where the breeze has a drying effect.

Many people love growing – and then eating – vegetables. They can head for the display of traditional and ornamental vegetables, which are harvested during the quarterly Gardeners' Day Out. One of the most interesting innovations is a garden devoted to plants native to Singapore, attracting native birds and insects and so helping sustain biodiversity.

The conservation-minded are able to pick up recycled gardening books, magazines or other equipment. The thrifty gardener can take advantage of the once-a-month distribution of free plant cuttings.

Pottering around

In the Pitter Patter Potter Garden (far left), children aged 5–10 can meet Mr and Mrs Potter (left). Here, they can enjoy themselves while developing their plant knowledge. The garden complements the primary school social- and natural-science curricula.

● Opposite: The Silver Garden, featuring a collection of grey, silver and white plants.

● Top: Visits to HortPark are very popular with school groups.

● Centre: Children learn about seed dispersal methods (left) and try their hand at harvesting (right).

● Right: Plants, gardening products and services available at the Gardeners' Day Out.

5
WORLDS OF NATURE

Singapore is small and its population is growing. Nonetheless, planners and park managers, with public support, have been able to conserve green areas representing a variety of natural habitats, filled with plant and animal life. These green spaces are distributed across the island – no one is far from a spot where one can look at nature, draw breath and de-stress. Nature's capacity to sustain itself has benefited from active management.

● A quiet track near Bukit Timah Hill.

Nature in the neighbourhood

Natural habitats are essential to wildlife. Green places also provide people with an escape from urban life. Every Singaporean, whether or not a car owner, lives within an hour or so's travel from a piece of forest, rich with greenery and abundant in animal life. To be sure, most of the primary forest has gone. However, many areas are being reforested with faster-growing secondary vegetation, among which native trees can be planted and allowed to mature in their own good time.

The law protects wildlife: the Parks and Trees Act covers land managed by NParks, including the nature reserves; the Wild Animals and Birds Act, administered by the Agro-Veterinary Authority (AVA), deals with wildlife in other areas.

In addition to the four gazetted nature reserves, Singapore also has a number of designated nature areas. Planners intend to preserve these for as long as possible. Some of them are surrounded by parks, acting as a buffer between them and the urban environment. One such area is Sungei Cina, embedded within Admiralty Park. And then there are the hundreds of parks themselves, and the park connectors, painting the island with splashes of green and offering a home to birds, butterflies, dragonflies and many other animals.

"I saw a hornbill today near my balcony at Gerald Drive. It was only 10 metres away. It was a beautiful sight. Seeing it perched on an Alstonia tree and cocking its head made my day."
"Invicta", a reader of *My Green Space*, NParks' online newsletter

● Hindhede Nature Park, created as a buffer between the Bukit Timah Nature Reserve and urban development.

Four nature reserves protected under the Parks and Trees Act

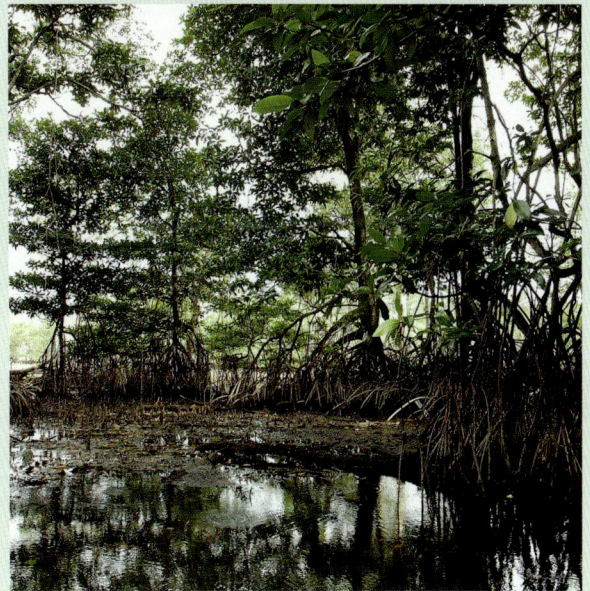

Top left: **Central Catchment Nature Reserve** consists mostly of secondary forest, but it includes some lowland Dipterocarp forest and freshwater swamp.

Lower left: **Labrador Nature Reserve** is a haven for intertidal biodiversity, the life that exists on the rocky seashore between the high-tide and low-tide marks.

Top right: **Bukit Timah Nature Reserve** is the best place to see lowland Dipterocarp forest.

Lower right: **Sungei Buloh Wetland Reserve** boasts mangroves, mudflats and grasslands.

Two parks at Bukit Batok

As Bukit Batok Nature Park shows, it may be possible for a forest to regenerate. In the early 20th century, this was the site of a granite quarry. The land around the workings was largely bare of vegetation. In the years since, trees have grown up to form a dense secondary forest, perfect surroundings for nature walks. The hilly terrain makes for good exercise, and there are high lookout points on the trails, giving spectacular views. The old quarry is now a lake. Around it, exposed to the elements, the granite bedrock of central Singapore forms fluted cliff faces and ominous overhangs stained red with iron oxide. Not far away is another lake, also formed from an old quarry, which forms the centrepiece of Bukit Batok Town Park and has been compared to Guilin in China.

During the Occupation, the Japanese military built a war memorial for their soldiers on Bukit Batok. The only traces today are a pair of pillars, 120 concrete steps and a memorial plaque.

"It's good that at some places such as Bukit Batok and Kent Ridge, the parks maintenance people have left part of the parks as they are, without over-cutting the grass. These two parks have a higher density of bird life compared to the more manicured parks."
Alan OwYong, past chairman of the NSS bird group

"My wife and I love how the l[...]re Park gives it a very natural charac[...]arks. Plus, we get to meet our friends h[...]se session. It's a great get-together pla[...]

David Wong, marketing manager

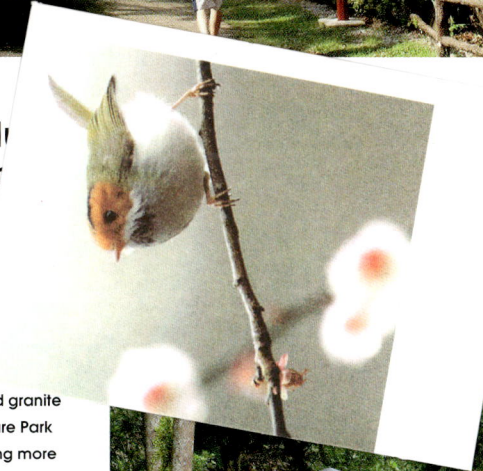

● Opposite, below: The granite formations at Bukit Batok Town Park look a little like the much bigger limestone formations at Guilin in China, hence the nickname "Little Guilin".

● Above: The exposed granite cliffs at Bukit Batok Nature Park are the result of quarrying more than half a century ago.

● Right: Bukit Batok Nature Park is an ideal spot for fitness enthusiasts.

The buffer zones

There was a time when the Dairy Farm and Hindhede Nature Parks were zoned for residential development. When the flora and fauna of Bukit Timah and the Central Catchment Nature Reserve were being surveyed in the early 1990s, it became clear that the reserves needed to be protected by buffer areas at their margins – both to insulate the natural habitat from surrounding urbanisation and to prevent encroachment by developers. The effect of concrete buildings extends beyond their immediate footprint – concrete reflects heat and affects humidity levels; high-rise blocks can have a wind-tunnel effect and cause damage to the nearby vegetation.

Through thoughtful deliberation among URA, NParks and other agencies, the Dairy Farm and Hindhede parks were designated "Nature Areas" under the Parks and Waterbodies Plan 2002.

The name of the Dairy Farm Nature Park comes from a farm that the Singapore Cold Storage Company set up in 1929 for the production of fresh milk. Within the park is the Singapore Quarry. This and the Hindhede Quarry, not far to the south, are now lakes surrounded by secondary forest, including fruit trees such as durian, rambutan, jackfruit and mango trees, and many other plant species. This is a favourite spot for rock climbers.

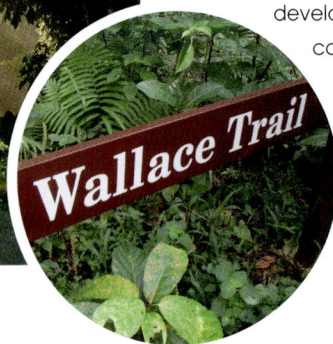

● Above: A relaxing walk at Dairy Farm Nature Park.

Teeming with wildlife

Dairy Farm Nature Park features information boards with illustrated extracts from writings by Alfred Russel Wallace, a contemporary of Charles Darwin who independently reached broadly similar conclusions on natural selection. In *The Malay Archipelago* (1869), Wallace wrote of his findings in this part of Singapore: "In about two months I obtained no less than 700 species of beetles, a large proportion of which were quite new, and among them were 130 distinct kinds of the elegant Longicorns (Cerambycidae), so much esteemed by collectors. Almost all of these were collected in one patch of jungle, not more than a square mile in extent, and in all my subsequent travels in the East I rarely if ever met with so productive a spot." Two beetles Wallace may have come across in his travels are a longhorn beetle of the genus *Oberea* (below left) and a Leaf-rolling Weevil (*Korotyaevirhinus necopinus orientalis*) (below right), also known as the Giraffe Weevil for its long neck.

Above: Cyclists keep to the track at Dairy Farm Nature Park.

Below left: The Wallace Education Centre is housed inside a restored former cowshed at Dairy Farm Nature Park.

Below right: The Wallace Environmental Learning Lab (WELL), established by NParks in collaboration with Raffles Girls' School and sponsored by Glaxo SmithKline and the Economic Development Board.

● Secondary forest surrounds and protects the Upper Peirce Reservoir.

The Central Catchment Nature Reserve

Since the founding of modern Singapore, maintaining the water supply has been a priority, which is one reason why the Central Catchment area has been spared redevelopment. For the managers of Singapore's water supply today, this protected status means less pollution and better water quality. There is a further benefit, as Chew Men Leong, chief executive of PUB, points out. "The rich vegetation and absence of hard surfaces means that the water flow is not subject to the great fluctuations in volume after heavy rain that can cause flooding problems for us elsewhere." Here, rainfall feeds a series of reservoirs – MacRitchie, Lower Peirce, Upper Peirce and Upper Seletar. An "impounding reservoir" was first completed near Thomson Road in 1868; in 1891, it was enlarged and renamed after municipal engineer James MacRitchie.

In 1990, the newly formed NParks took charge of the nature reserves. Up to then they were largely unmanaged, apart from basic maintenance. NParks undertook a biological survey of the Central Catchment area as a basis for future conservation planning. The part of the survey devoted to trees was carried out under former parks commissioner Wong Yew Kwan and resulted in a useful vegetation map. It showed there was still some primary forest outside the Bukit Timah reserve. Animal surveys followed.

By 1995, enough data had been gathered to form the basis of the Nature Reserves Recreational Master Plan, which identified areas of high biodiversity and fringe areas more suitable for public facilities. NParks was able to make the case for areas to be approved as buffers for the reserves, which is how the Hindhede, Singapore Quarry and Dairy Farm nature areas came about. All these efforts, including surveys, documentation and improved conservation management, were only possible because NParks had the collaboration of many partners – individuals, academics, non-governmental organisations (NGOs), corporations and other public agencies.

Forest glider

The Malayan Colugo (*Galeopterus variegatus*) is a little-known tree dweller that leaps through the forest with the aid of a special kite-shaped membrane, known as a patagium. It can glide over distances of 100 metres or more. It feeds on young leaves, shoots, flowers and fruit. After giving birth, a mother will nurse her baby for about six months, wrapping the patagium around it to form a protective pouch.

Return of a tiny forest-dweller

The mousedeer is back! In 2008, a Greater Mousedeer was recorded on Pulau Ubin by an automatically triggered camera set up by a volunteer working with NParks on a wildlife survey. This was the first official sighting in more than 80 years. Others followed. Although fairly common in the region, the Greater Mousedeer (*Tragulus napu*) was thought to be extinct in Singapore; the Lesser Mousedeer (*Tragulus kanchil*) is critically endangered. These shy creatures are among the smallest hoofed animals in the world, with a head-and-body length of just 52–57 cm (Greater) and 40–50 cm (Lesser), and weighing no more than 4.5 and 2.5 kg respectively. Mousedeer have also been seen around the Lower Peirce Reservoir boardwalk and in the Bukit Timah Nature Reserve. The Greater Mousedeer sometimes hides from predators by walking into water, where it can stay submerged for up to five minutes.

Monkeys in the forest

All but a few of Singapore's monkeys are Long-tailed Macaques (*Macaca fascicularis*). Sir Stamford Raffles gave this species its scientific name in 1821. In Singapore, there are thought to be around 1,500 in all, living in troops of up to 30 in the nature reserves and other densely vegetated areas. It spends most of its time up in the branches of trees, but occasionally forages on the ground. It feeds mostly on fruits and seeds, occasionally preying on small animals. It has an ecological role, helping to regenerate the forests by eating and dispersing seeds. Long-tailed Macaques are highly intelligent. Unfortunately, people persist in feeding them, despite signs making it clear that feeding monkeys is an offence.

A tree-dweller threatened with extinction

The Banded Leaf Monkey (*Presbytis femoralis*) is a timid creature found in Singapore and Johor. Although it is one of the island's rarest mammals, with a population estimated at around 40, this is an improvement over its low point in the 1970s, when there were probably only a dozen. Some of the plants it likes to feed on are also themselves threatened, demonstrating the complex challenges involved in maintaining biodiversity. The common name comes from the white line that runs up the front of its body; the hair on its belly is otherwise grey, and the hair on its back is black.

The biological survey of the early 1990s enabled NParks to establish the core conservation areas within the Central Catchment Nature Reserve. After that, it was possible to plan the trail network. It took three people some 18 months in 1993–94 to draft a baseline map, starting at MacRitchie and moving on to Nee Soon and Mandai. One member of the team was Sharon Chan, who later managed the Sungei Buloh Wetland Reserve. "It was challenging work – we could only do half a day at a time!" Many existing small trails, some of them dead ends, had to be closed off. By 1996, the map was finalised.

The first phase of the Recreational Master Plan for the Nature Reserves was implemented by 2001; by 2004, the picture was almost complete, with all the new facilities up and running. Today, there are marked trails around the MacRitchie Reservoir, ranging from 3 km to 11 km in length and from "easy" to "difficult". There are four boardwalks along the water's edge.

Nature thrives most when left undisturbed. At the same time, people want to see it close up and learn about it. The Central Catchment area shows how, with imagination, these two requirements can be reconciled. And Singapore gets a supply of clean water as part of the bargain.

● Top left: The Red-crowned Barbet (*Megalaima rafflesii*) lives in the canopy of lowland rainforests. It nests in cavities excavated in tree trunks.

● Top right: The Chestnut-bellied Malkoha (*Phaenicophaeus sumatranus*) is a species of cuckoo native to Southeast Asia but threatened by habitat loss.

● Opposite: The TreeTop Walk is a 250-metre-long suspension bridge between the two highest points in the Central Catchment Nature Reserve.

"When I was in the army, although we were often rushing from point to point, trying to meet 'H-Hour' for some mission or other, it was often quite an experience when we took a breather and observed the world around us. It could be in the middle of the night, standing by a stream, listening to the many insects singing away as soldiers trudged across the waist-deep waters; or at dawn, catching the morning sun peering through the lush trees, as we sat down on a small hill. I would be tired but always amazed by a part of Singapore seldom seen."

Tan Chuan-Jin, senior minister of state for National Development

Isolated, inaccessible and fascinating
Within the Central Catchment Nature Reserve is the Nee Soon freshwater swamp forest. This is the only remaining sizeable patch of the primary swamp habitat that was more extensive two centuries ago. Here live some animal species that are found nowhere else in the world. It is only accessible with special permission. However, for Professor Peter Ng, head of the Raffles Museum of Biodiversity Research at National University of Singapore, this is one of the nation's most interesting natural environments. "The swamp forest is isolated and a very wet place. It's home to a very diverse range of freshwater crabs and fish, arguably the most valuable freshwater biodiversity hotspot on the island."

Living with the military
The army still uses part of the Central Catchment area for training, but it tries to be as unobtrusive as possible. The Ranger Station near the TreeTop Walk was once a depot for three-tonner trucks during exercises. Drawing up the Recreational Master Plan involved negotiation. As Sharon Chan recalled, "We needed to know the no-go areas from the point of view of Mindef, as well as PUB. The Nee Soon swamp area is a non-recreational zone – it is a live-firing area. If we want to do research in the area we have to get prior clearance. Without clearance, it is out of bounds."

Pulau Ubin

Once a cluster of smaller islands separated by tidal rivers, Pulau Ubin ("Granite Island") was transformed into a single island by the building of bunds for prawn farming. It is home to a rich variety of plants, mammals, reptiles, amphibians, butterflies and birds.

In earlier times, granite quarrying supported several thousand settlers. Today, the granite quarries are abandoned, but there are still a few residents – the island is an interesting reminder of the way many Singaporeans lived in the kampong era. Some villagers still engage in traditional farming and fishing. In contrast to modern, mainland Singapore, the island relies on wells for its water supply, and diesel generators for electricity.

Under URA's Parks and Waterbodies Plan, the planning vision is for Pulau Ubin to be kept in its rustic and natural state for as long as possible so that it can serve as an outdoor playground for Singaporeans to enjoy.

Pulau Ubin attracts a regular flow of Singaporeans and tourists in search of a quiet island getaway. For nature lovers, the discovery of exceptional marine biodiversity on the beach at Chek Jawa has added to the island's appeal.

● A beach at Pulau Ubin. Such scenes were once common around Singapore's coastline.

● Opposite, top: The locally endangered Buffy Fish Owl (*Ketupa ketupu*) lives in forests and mangroves and can be found not only on Pulau Ubin but also in the Central and Western Catchment areas, Sentosa and Loyang. It is endangered due to illegal trapping.

● Above left: The Red Junglefowl (*Gallus gallus*), a locally endangered species found in the wild on Pulau Ubin. It is believed to be the ancestor of the domesticated chicken.

● Above right: The Mangrove Pitta (*Pitta megarhyncha*) has disappeared from mainland Singapore due to loss of habitat. Only a small population remains on Pulau Ubin and Pulau Tekong.

Return of the hornbills

The Oriental Pied Hornbill (*Anthracoceros albirostris convexus*), native to Singapore, disappeared from the forests in the mid-1800s, probably due to hunting and loss of habitat. It needs mature trees with natural cavities in which to make a nest. A breeding pair was sighted on Pulau Ubin in 1994, and now the species is making a comeback. As part of the Singapore Hornbill Project, artificial nestboxes were provided, and the hornbills began to re-establish themselves in various parts of Singapore. It is estimated that there are now over 100 resident birds.

The Singapore Hornbill Project is a collaboration between NParks, Nanyang Technological University, the National University of Singapore, Wildlife Reserves Singapore (which oversees Jurong BirdPark) and other sponsors.

In 2008, one captive-bred pair was introduced to a site in the Bukit Timah area, and with the aid of a temporary aviary and artificial nest, produced chicks that fledged successfully. NParks conservation director, Wong Tuan Wah, was a happy man. "My joy and relief were comparable to that of a newborn baby's parents!"

Pulau Ubin from above

A composite aerial photograph shows the varied face of Pulau Ubin. The granite quarries and prawn farms, which were once the basis of the island's economy, are clearly visible. Today, Pulau Ubin is a peaceful rural getaway, where visitors can get close to nature.

Ketam Quarry

Kekek Quarry

Outward Bound Singapore

Ketam Mountain Bike Park

Pulau Ketam

Ubin Quarry

View from Bukit Puaka (Pulau Ubin's highest point)

Try spotting these animals when you visit

Greater Mousedeer
(*Tragulus napu*)

Green Crested Lizard
(*Bronchocela cristatella*)

Sunda Pangolin
(*Manis javanica*)

Paradise Tree Snake
(*Chrysopelea paradisi*)

Fish Farms

Mamam
Campsite

Floating Fish Farms

National Police
Cadet Corps
Campsite

Balai Quarry

Chek Jawa
Wetlands

Chek Jawa
Visitor Centre

Pulau Sekudu

Butterfly Hill

Jetty

Ubin-HSBC
Volunteer Hub

Nature Gallery

Pekan Quarry

Jelutong
Campsite

Pulau Sekudu

Tudor-style cottage

Chek Jawa

At the eastern end of Pulau Ubin, in 2000, a naturalist accompanying a school party discovered Tanjong Chek Jawa, a very interesting intertidal area where several coastal ecosystems meet in one place – sandy and rocky beaches, seagrass lagoon, coral rubble, mangroves and coastal forest. In response to plans to reclaim land at Chek Jawa, nature groups made a convincing case for its conservation. In December 2001, the government announced that reclamation would not go ahead so long as the land was not needed for development.

Ironically, Chek Jawa's natural heritage was soon threatened by the pressure of visitor numbers and a spell of exceptional rainfall. NParks, Nature Society (Singapore) (NSS), the Raffles Museum of Biodiversity Research and the Singapore Environment Council worked together to create a management plan and to monitor the ecological health of the site.

In 2007, the public were admitted to a set of facilities, including a visitor centre with viewing jetty, a kilometre-long boardwalk and a viewing tower. Now visitors are less likely to impact the biodiversity they have come to see.

● Top left: A rocky beach, one of several coastal ecosystems at Chek Jawa.

● Centre left: The Estuarine Seahorse (*Hippocampus kuda*) anchors itself with its prehensile tail.

● Below left: A dense cluster of Common Button Tops (*Umbonium vestiarium*). The population of this mollusc has declined through over-collection.

● Opposite: On the intertidal foreshore, a Carpet Anemone (*Stichodactyla* sp.).

Labrador Nature Reserve

Labrador, which became a gazetted nature reserve in 2002, is important for the life along its rocky shore. The intertidal zone (the area on the beach between the high and low tide marks, to the east of the pier) has a variety of algae, snails and other molluscs, crustaceans and seagrasses. A narrow strip of coral here is the only coral community on the shore of mainland Singapore.

The reserve offers a good view of the South China Sea. Backing it is a rocky sea cliff, surmounted by a coastal forest – walking trails lead into the cool, dark interior, and lookout points give panoramic views over the sea. The forest is home to over 80 kinds of birds and many species of butterflies, spiders and insects. The height of the cliff made Labrador a good military defence position during colonial times, and relics of wartime gun emplacements can still be seen buried in the forest.

● The Rufous Woodpecker (*Celeus brachyurus*), found in the wild in Southeast Asia and the Indian subcontinent, can be seen at Labrador Nature Reserve.

"The Labrador Nature Reserve has a living, instructional beach of great scientific and educational value. The species diversity is extraordinary, considering the numerous pressures the 300-metre stretch of rocky shore faces. Almost all the major animal phyla (taxonomic groups) can be found at Labrador."

Prof Leo Tan, chairman, Garden City Fund

● Left, above: Gun barrel used for coastal defence, with a range of nearly 10 km. Similar guns were deployed by the Labrador Battery during the Japanese onslaught in February 1942.

● Left, below: The Berlayer Creek and Bukit Chermin boardwalk is outside the gazetted reserve at Labrador but is a great vantage point for viewing wildlife and mangroves.

● Below: A family looks out to sea from the Bukit Chermin Boardwalk. Labrador Nature Reserve is in the background.

A view of the sea from a lookout point in Labrador Nature Reserve. The jetty and rocky foreshore are inaccessible to the public, in order to protect the natural habitat.

Sungei Buloh Wetland Reserve

● Above: The East Asian-Australasian Flyway.

● Inset: Nordmann's Greenshank (*Tringa guttifer*), a rare visitor that breeds on Sakhalin Island and in the west Okhotsk Sea.

Two hundred years ago, much of the northern coastline would have looked like Sungei Buloh Wetland Reserve, made up of mudflats, river estuaries and tidal creeks. Until recently, prawn and fish farming were the main human activities here. In 1986, the area was discovered by some keen birdwatchers from the Singapore branch of the Malayan Nature Society, predecessor of the NSS. After the society and other interested parties approached the government, it was designated a nature park and developed with input from the Wildfowl and Wetlands Trust in the United Kingdom (founded by the ornithologist Sir Peter Scott) and the Worldwide Fund for Nature. It was declared officially open in 1993 by the then prime minister Goh Chok Tong. The following year, the reserve welcomed its 100,000th visitor. In January 2002, it became a gazetted nature reserve, and two years after that, Singapore's first ASEAN Heritage Park.

Out of 62 species of mangroves still to be seen in Singapore, 53 can be found in Sungei Buloh. They provide a habitat for over 2,000 species of other plants and animals, from tiny crabs to Malayan Water Monitors.

The bird life here is world-famous. The reserve is a pit stop on a bird migration route known as the East

● Opposite, above: Aerial view of Sungei Buloh Wetland Reserve's visitor centre.

● Above: Arrival of a flock of Whimbrels (*Numenius phaeopus*), a common visitor that breeds in central Siberia.

● Below: Mangrove plants have roots that absorb water and nutrients while excluding salt and can grow in brackish water and along the seashore.

● Inset below: Crabs and other creatures make their homes on the roots and branches of the mangroves.

Asian-Australasian Flyway, extending from Alaska, eastern Russia, China and Japan to Australia and New Zealand. The most numerous species are plovers, sandpipers, herons and egrets. Over 230 bird species have been recorded in Sungei Buloh, of which 118 species are migratory. Many of the wading birds feed on the nearby mudflats of Mandai and Kranji, but fly into the reserve when the tide is in to roost as well as to feed.

The boardwalks and birdwatching hides give visitors one of the best opportunities to observe nature in Singapore, particularly in the early part of the day. Sharp-eyed observers who are able to watch patiently and quietly get to see the most.

Singapore's biggest lizard

The Malayan Water Monitor (*Varanus salvator*), not to be confused with the Estuarine Crocodile (*Crocodylus porosus*), is a common sight at Sungei Buloh Wetland Reserve, in mangroves, forests, scrubland and even next to urban canals. An able climber and graceful swimmer, it is one of the largest lizards in the world, capable of growing up to 3 metres in length. It uses its forked tongue to detect odours in the air as it searches for prey.

On the fiddle

A close look at a stretch of natural shoreline, particularly near mangroves, will probably reveal a host of busy little Fiddler Crabs (known to scientists as members of the genus *Uca*), with bodies just 2–3 cm across. The male has one huge pincer, often as large as the rest of the body! It is used to attract females and deter rivals. He waves it from side to side rather like the right hand of a violinist, hence the name "fiddler". The other pincer is used for feeding. Females have two smaller pincers.

A romp of otters

A young otter is known as a "pup", and a family is a "romp". Several Smooth Otters (*Lutrogale perspicillata*) have made their home at Sungei Buloh Wetland Reserve, where they have been eating, breeding, marking territory or just playing. One of their favourite places is Sungei Buloh Besar, where they hunt for fish when the tide is coming in. Their streamlined shape and large webbed feet make them graceful and effective swimmers. Photographers have commented on their apparent eagerness to show their faces to the cameras! For couch potatoes, they can watch these animals' antics online at www.nparks.gov.sg/animalcam.

The Southern Ridges

In 2008, three nature areas, Mount Faber Park, Telok Blangah Hill Park and Kent Ridge Park were joined together to form the Southern Ridges.

Mount Faber was named after Captain Charles Faber of the Madras Engineers, who constructed a road to the top so that a signal station could be set up for communication with ships. Because of its height, the hill is now a boarding point for the cable car crossing to Sentosa.

The land occupied by Kent Ridge Park was originally forest with rubber and pineapple plantations – some rubber trees still remain. Planning for the park started in the late 1970s, with the intention of maintaining the natural features of the ridge. By 1988,

two years after the National University of Singapore campus was built nearby, the park was completed. There are still groves of Tembusu, Acacia (*Acacia auriculiformis*) and Simpoh Air (*Dillenia suffruticosa*) trees, and a nature trail with pitcher plants and wild orchids.

Kent Ridge (so named in 1952) was the setting for fierce fighting during the Japanese invasion. In the last-ditch Battle of Pasir Panjang, fought in February 1942, a remnant of the Malay Regiment, having run out of ammunition, fought hand-to-hand with the 18th Division of the Imperial Japanese Army. Their courage is commemorated in a museum, Reflections at Bukit Chandu ("Opium Hill", a reference to a former

opium factory that was close by).

The 9-km walking trail runs all the way along the Southern Ridges from HarbourFront to Clementi. The route includes two ultra-modern bridges – Alexandra Arch and Henderson Waves – together with two elevated walkways, the Canopy Walk at Kent Ridge Park and the Forest Walk at Telok Blangah Hill Park. The terrain is hilly and varied, giving some of the best views over the city and out towards the sea.

When these parks were isolated from one another, there were relatively few users. Now the Southern Ridges, with three parks seamlessly integrated, has become a favourite weekend destination.

Planting for the future

Over 600 trees on the heights of Mount Faber and Telok Blangah Hill are being incorporated into a new arboretum, the SembCorp Forest of Giants, made possible by SembCorp. These species once dominated the regional landscape. In the wild some of them can reach a height of over 80 metres. The collection will help researchers identify new species for planting along roadsides and in parks.

The Singing Forest, supported by ST Microelectronics, is a collection of regional native trees that provide food sources, shelter and nesting areas for native birds.

● Opposite: Henderson Waves, the highest pedestrian bridge in Singapore, offers a spectacular view of the Southern Ridges.

● Above: The view from Mount Faber.

● Below: The 800-metre Forest Walk provides visitors with an elevated view of the secondary forest of Telok Blangah Hill.

The right balance

The development of visitor facilities in the nature reserves was the product of a "recreational master plan", which was a follow-up to a comprehensive biological survey. The thinking was that while NParks has the capacity to conserve nature, public backing is also essential. Peggy Chong, now deputy chief executive officer of Gardens by the Bay, previously managed several key NParks development projects. She said,

"To justify the protection of nature, the public has to be exposed to it. By allowing some form of recreation, you get more public support." Once their eyes were opened, many people became fervent champions of nature conservation and acted as extra eyes and ears for NParks.

At the same time, feedback showed the need for public education. Some things are appropriate to a nature reserve, other things are not. Some visitors asked for concrete paving on the paths and trails in nature reserves and nature areas. They complained that their shoes got dirty or that it was difficult to walk the trails in high heels. Conservation officers explained that the idea behind a facility such as the TreeTop Walk is to show the forest as far as possible in its pristine state. Paving would spoil that. Others wanted the nature reserves and their carparks to be lit up at night. This

would disrupt the wildlife, especially the nocturnal animals – furthermore, NParks discourages visits to the reserves at night for the safety of both people and animals.

There are practical constraints too. There have been requests for a cable car to the top of Bukit Timah Hill. The installation of electricity and water supplies for visitor facilities would destroy the effectiveness of the hill as a nature reserve. In remote parts of Pulau Ubin, there is no piped water or electricity, which makes the installation of sophisticated toilets and showers impossible.

Many people like parks and reserves to be relatively unmanaged, to maintain their rustic appeal. They like to experience a sense of adventure in a safe environment. This is why NParks has provided a broad range of parks and reserves, with various degrees of "naturalness".

● Opposite, top: Shelters at Sungei Buloh Wetland Reserve provide visitors with a place to rest as well as observe the wildlife.

● Opposite, below: The eco-toilet at Tampines Eco Green turns human waste into compost for the plants.

● Above left: A guided nature tour for students.

● Above right: The Aerie is an 18-metre-high tower, giving a panoramic view of Sungei Buloh Wetland Reserve.

● Below: The boardwalk at MacRitchie Reservoir offers an opportunity to study the mudflat denizens close-up without the need to get muddy.

Respecting the nature reserves

As public interest in nature grows and education becomes more effective, people show greater responsibility in their use of the parks and reserves. Visitors get the idea that if they crash around or make a lot of noise, animals will generally do everything they can to hide. Even so, NParks still has to communicate some basic messages. For example, dogs are not allowed in the nature reserves for a good reason – they disturb and sometimes even kill animals. Plants and animals should not be abandoned in the parks or nature reserves – in fact, most domesticated animals, if released into the wild, will not live for very long.

Singapore's forests are very dense, and even experienced guides can get lost. In September 2012, according to a *Straits Times* report, two marathon runners strayed from the marked trail between Upper Peirce and Upper Seletar reservoirs. As night was falling, they realised they had lost their bearings. After reporting their location by mobile phone, they were rescued by boat at 8.50 p.m. As one of them said, "To continue to bash through the night could have been dangerous. There could have been wild animals, snakes, or we could have lost our footing and got injured." Even on Pulau Ubin, generally thought to be very safe, cyclists have been injured because they were not prepared for the terrain.

● Left: A Laced Woodpecker (*Picus vittatus*) feeding its chick.

● Right: The Ketam Mountain Bike Park on Pulau Ubin.

Advantages of a natural diet

Animals' natural feeding habits are part of the ecosystem. For example, monkeys consume fruit containing seeds. By expelling the seeds as they move around, the monkeys play a part in dispersing them, helping plants to multiply. Feeding the monkeys disrupts this pattern. Having been fed once, monkeys expect the same thing to happen again, and this can lead to aggressive behaviour. NParks has put on roadshows, workshops, guided walks and other outreach programmes to get this message across.

Better flying free

The Oriental Magpie Robin (*Copsychus saularis*) is a distinctive little bird with black-and-white plumage. It was once one of the most common birds in Singapore's gardens. It lives on a diet mostly of worms, grasshoppers and other insects. Oriental Magpie Robins are great singers and can even imitate the calls of other species. This is a protected species in Singapore – from the 1960s onwards, numbers were reduced through destruction of their natural habitat, competition from mynas and trapping for the cagebird trade. Sadly, they do not live for long in captivity. By the early 1980s, there were less than 50 birds countrywide. Efforts to reintroduce them have met with some success. A few individuals can be seen in quiet locations such as Pulau Ubin and Sungei Buloh.

Unintentional cruelty

Animals bought from markets or dealers often end up unwanted. Non-native species released into the wild often die. For example, in July 2012, a Wattle-necked Softshell Turtle (*Palea steindachneri*) was found in Tyersall Avenue with a cracked shell. A native of China, it was thought to have been imported illegally. It was rescued, but later died of its injuries. While cultural traditions can be respected, releasing caged birds or commercially bought terrapins into the wild, as used to be done on Vesak Day, can have cruel consequences (and is now prohibited by law).

In 2005, NParks and PUB, together with partners and volunteers, embarked on Operation No Release, an annual campaign to raise awareness during the run-up to Vesak Day about the consequences of releasing animals.

How close to nature do you want to be?

In kampong days, up-close contact with birds and other animals was an accepted part of daily life. For most people in the 21st century, it is not. As Singapore gradually evolves into a City in a Garden, the question arises, "How 'natural' should the garden really be?"

Close encounters between modern living and nature have not always been happy ones. When residential developments were built closer and closer to wooded areas, food scraps disposed of in litterbins, or left out near open doors and windows, attracted monkeys. Some birds, particularly crows, pigeons and mynas, used to thrive in the days of open-air hawker stalls and rubbish spilling out of dustbins. Now, more food courts are becoming enclosed, and flat dwellers drop their rubbish down enclosed chutes. The birds are less of a nuisance than they were, but they still can cause annoyance and mess. The trees planted in Orchard Road look beautiful, but pedestrians have not always welcomed the thousands of starlings and mynas that roost there in the evening and sometimes leave their mark. In open carparks, bird droppings irritate the owners of nice shiny cars.

Living in a garden city requires some tolerance of nature. As Dr Cheong Koon Hean, CEO of HDB, put it, "Ecological corridors are something we support in principle. But you have to accept that you will have monkeys, squirrels and bird droppings. You can't have a shady tree and ask for the tree not to shed leaves."

● Below left: Squirrels are a familiar sight close to houses and flats, and in parks.

● Below right: An Olive-backed Sunbird (*Nectarinia jugularis*), whose natural habitat is mangroves, but is often seen in urban surroundings.

Animals such as pigeons (top), mynas (above right) and monkeys (left) are easily attracted by food remnants carelessly left in the open. The birds can dirty floors with their droppings, while monkeys can be harmed by containers such as cans and plastic bottles.

6

BIODIVERSITY, A SHARED HERITAGE

Despite development, nature surveys have shown that Singapore still has a wide variety of biodiversity. Coordinated effort and expertise have laid the foundations for its long-term conservation. Year after year, new species have been discovered; others that were feared lost have been rediscovered. The variety of flora and fauna fascinates young and old alike. Public awareness will ensure that this vast gene bank will remain one of Singapore's prime resources.

● Many organisms grow on trees, which provide habitats for a variety of plants and animals.

Richness in diversity

When the PRD first set about creating a garden city, no one talked about biodiversity. Ecology was left to academic botanists and zoologists. Over the years that followed, attitudes changed. Helping to maintain the balance of nature was always the job of nature reserves. Over time it became part of the role of man-made gardens and ornamental greenery too.

Does biodiversity matter? It maintains our sources of nutrition, fresh water and fuel. It helps stabilise temperature and air quality. It is an important source of raw material for new medicines as well as a range of industrial processes. It is an educational resource.

Once we lose biodiversity and allow the gene pool to be reduced, regeneration is slow and difficult. This is a particular challenge in Singapore. The trees of the primary forests are slow-growing. Habitats have been fragmented by development. Flora and fauna need good nutrition, water and air; they are sensitive to toxins just as humans are. Biodiversity is the proverbial "canary in the coal mine" – when it shows signs of stress, humans should heed the warning signs.

NParks is developing green corridors that are known as "Nature Ways", linking existing parks and nature reserves to encourage birds and butterflies to move between them. This will greatly increase the amount of natural habitat available to them. The first link will be the Kheam Hock Nature Way. Others will follow at Admiralty, Tampines and Yishun. The complete network will be some 60 km long. The public, including schoolchildren, will be able to help in the planting and nurturing of greenery in these new "linear parks".

"Protection of biodiversity is important, especially since it is disappearing at an unprecedented rate. Singapore is a good case study in illustrating how economic development and biodiversity conservation can be mutually reinforcing."

Tan Chuan-Jin, senior minister of state for National Development

● Pioneer plant species take root on previously cleared land.

The little red dot

The Crimson Sunbird (*Aethopyga siparaja*) won an informal poll organised by the NSS in 2002, to become Singapore's "national bird". Sunbirds have long, downward-curving bills that enable them to probe flowers in search of nectar. Of all the birds commonly seen at Sungei Buloh Wetland Reserve, the tiny sunbirds are among the most colourful. Six species occur in Singapore and include (clockwise from top left): the Crimson, Olive-backed (*Nectarinia jugularis*), Purple-throated (*Nectarinia sperata*), Copper-throated (*Nectarinia calcostetha*) and Brown-throated (*Anthreptes malacensis*) Sunbirds.

Butterflies in the city

A profusion of butterflies is a sign that nature can flourish – even in the city centre. The NSS initiated a project to establish a series of "hotspots" with collections of butterfly-friendly plants, linked in such a way as to make up a larger, integrated habitat. The Butterfly Trail@Orchard runs from the Singapore Botanic Gardens at one end, via Cuscaden Road, Orchard Boulevard, Somerset Road and Orchard Road to Fort Canning Park. Within three years of its establishment in 2010, the trail has attracted more than 60 species of butterflies, including the Leopard Lacewing (*Cethosia cyane*) (right) and the Five Bar Swordtail (*Pathysa antipathes itamputi*) (far right).

Biodiversity in numbers

There are literally millions of plant and animal species in the world – some say tens of millions. Around 1.8 million have formal scientific names, but botanists and zoologists believe there are many more waiting to be discovered. In Singapore, the named species are certainly not the whole story. The island is located right in the centre of Southeast Asia, which has been described as "the most important biodiversity hotspot on the planet". The confluence of the Indian and Pacific Oceans results in high marine biodiversity.

NParks has taken a number of active steps to re-establish biodiversity. For example, the staff at the Pasir Panjang Nursery succeeded in propagating the Singapore Kopsia (*Kopsia singapurensis*), a tree that is critically endangered in Singapore. In all, the nursery has grown more than 160 species of native trees and shrubs. The Singapore Botanic Gardens does similar work.

Many groups of organisms have not been studied in detail; there are regular new discoveries. Nearly 500 species of plants, animals and algae new to Singapore were recorded over the ten years up to 2012. Of these, more than 100 were new to science. Twenty-six species of plants and animals previously thought to be extinct in Singapore have since been rediscovered.

"Our heritage of animals and plants is something we can boast about. But if we don't conserve what we have now, we may have already lost too much, even if Singapore's drive for sustainable living succeeds. It may well be too late then!"
Prof Peter Ng, director, Raffles Museum of Biodiversity Research, National University of Singapore

● The Changeable Lizard (*Calotes versicolor*) was introduced into Singapore in the 1980s and has become widespread. In the breeding season, the male's head turns orange with a black blotch. Males display to females and rivals by doing push-ups.

The number of non-microbial species in Singapore has been estimated at over 40,000, an impressive tally for such a small country. The list below represents only those species that have been formally named.

vascular plants **3,971**
of which:
2,145 are native
1,826 are exotic

lichens **296**

mammals **52**

birds **364**

reptiles **103**

amphibians **28**

butterflies **301**

dragonflies **127**

reef fish **107**

freshwater fish **66**

gobies **149**

hard corals **255**

sponges ~ **200**

molluscs **1,264**

echinoderms **90**

seagrasses **12**

● From top: Changeable Hawk-eagle (*Spizaetus cirrhatus*), Oriental Whip Snake (*Ahaetulla prasina*), Milkfish (*Chanos chanos*) and a sponge (*Haliclona* sp.).

Discovered alive

The rediscovery of a species thought to be extinct in Singapore is an exciting moment. A Neptune's Cup (*Cliona patera*), a giant sponge, was recently found on the seabed off Singapore's southern coast. First identified in 1822 and last seen in the region a century ago, this sponge can grow so big that it was once used as a baby's bath!

The seabed at Pulau Pawai.

Sea anemones

Sea anemones are related to corals and jellyfish. When their tentacles are extended, they resemble flowers. Dr Daphne Fautin, from the University of Kansas, started studying Singapore's sea anemones in 2006. She found that there are twice as many species here as there are on the coast of North America between Vancouver and San Diego. New ones are still being discovered. Shown above, clockwise from top left are the Strawberry Sea Anemone (*Corynactis californica*), Striped Sand Anemone, Peacock Anemone and Frilly Sea Anemone (*Phymanthus* sp.).

The Rail Corridor

A chance to establish a continuous green link across Singapore, virtually coast to coast, arose when the terminus for the rail service from Malaysia was moved from Tanjong Pagar to Woodlands. In the absence of development in the past, nature has flourished beside long stretches of track, in a setting made up of forest, grassland, small farms, canals, streams and marshland. The route links many areas of great biodiversity and could well offset some habitat fragmentation. In 2011, URA started to study the idea of maintaining the corridor as a green link between the larger network of green spaces, park connectors, Nature Ways and the future Round Island Route. The public was invited to contribute ideas. SMS Tan Chuan-Jin said, "Development does not mean that we need to lose the unique attributes of the corridor, which are connectivity, greenery and a special heritage in the railway."

Research as a basis for action

Singapore has two organisations focused on biodiversity: the National Biodiversity Centre (NBC), which comes under NParks, and the Raffles Museum of Biodiversity Research (RMBR), part of the Faculty of Science at the National University of Singapore. One of the roles of the NBC is to bring together, in one place, all available information on Singapore's biodiversity, so that it can be used in official policy and decision-making. It also represents Singapore internationally in biodiversity-related forums. RMBR has a broader research and educational remit, with a regional, indeed global, perspective. The two organisations often work together.

The year 2009 saw the publication of "Conserving Our Biodiversity – Singapore's National Biodiversity Strategy and Action Plan". It reflects the objectives of the Convention on Biological Diversity, an international treaty signed by nearly 200 countries, including Singapore. Acknowledging the trade-offs involved in combining conservation with economic growth, it called for "unique solutions to create a nature conservation model that champions environmental sustainability in a small urban setting". Among other initiatives, it highlights the conservation and recovery of native species, including birds, dragonflies and rare plants, and points to the need for sufficient

habitats. The key thrusts are:
- Safeguarding biodiversity
- Considering biodiversity issues in policy- and decision-making
- Improving knowledge of biodiversity and the natural environment
- Enhancing education and public awareness
- Strengthening partnerships with all stakeholders and promoting international cooperation

● Above: Collecting data at the coral nursery off Pulau Semakau. Supported by Keppel Corporation, the nursery aims to grow coral from naturally broken fragments that can be transplanted onto coral reefs.

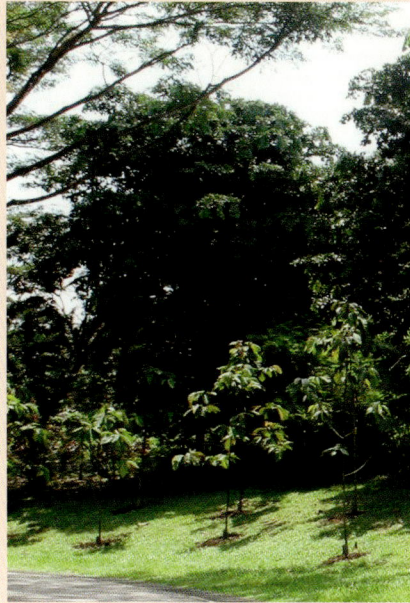

Yishun Arboretum

The primary forests of Singapore consist mostly of trees in the family Dipterocarpaceae – the Dipterocarps dominate the rainforests of Southeast Asia. The greatest diversity within the family is found on the island of Borneo. Their numbers have been drastically reduced throughout much of Southeast Asia because of land clearance and because they provide commercial products – timber, aromatic oils and resins. The Yishun Arboretum, which opened in 2008 with the support of Banyan Tree Holdings Ltd, is a living gallery of these trees and is designed for research and education. It contains over 800 trees, representing eight genera and 73 species.

Marine megasurvey

The Comprehensive Marine Biodiversity Survey is an effort to create a national database on Singapore's mudflats, intertidal areas, coral reefs and seabed. It involves a huge team of global and local experts, volunteers and enthusiasts, engaged in sampling, collecting, photographing, cleaning, sorting and cataloguing. The survey is supported by Asia Pacific Breweries, Care-for-Nature Trust Fund, Shell Companies in Singapore and the Air Liquide Group. Shown here are a feather star (left) and a flatworm (*Pseudoceros* sp.) (right).

Celebrating biodiversity

Singapore's first Festival of Biodiversity, held in May 2012, showed that the public – younger people especially – are hungry for information about wildlife and greenery. Over two days at the Singapore Botanic Gardens in May 2012, more than 40 partners took part, including individual volunteers, NGOs, nature and leisure interest groups, schools, companies and other organisations. The public were treated to symposiums, talks, book launches, a photographic exhibition, an exhibition on marine biodiversity and various fringe activities.

Singapore: birthplace of interesting species

Endemic (found only in Singapore)

Singapore Freshwater Crab (*Johora singaporensis*)

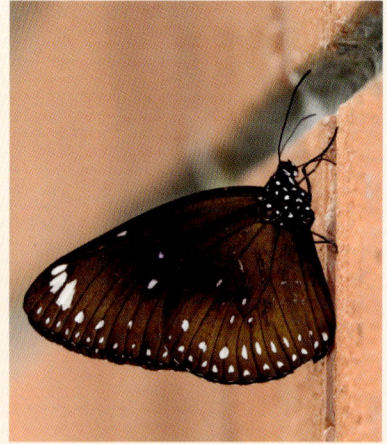

Blue Spotted Crow (*Euploea midamus singapura*)

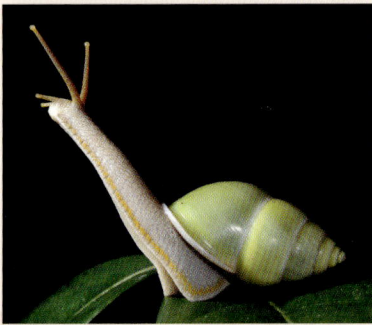

Green Tree Snail
(*Amphidromus atricallosus temasek*)

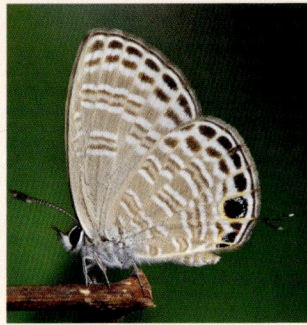

Singapore Fourline Blue
(*Nacaduba pavana singapura*)

Temasek Shrimp (*Caridina temasek*)

The Singapore Index on Cities' Biodiversity

Prior to 2007, there were only indices that ranked countries on their environmental sustainability efforts, concentrating on physical parameters like water supply, air pollution, carbon footprint, etc. As Singapore is a small city-state, it was excluded due to its size. Moreover, biodiversity indicators were often not included. To address this gap, Singapore proposed the development of a self-assessment tool for cities to evaluate their biodiversity efforts.

In 2008, the then minister for National Development, Mah Bow Tan, backed the idea. It was supported by the then executive secretary of the United Nations Convention on Biological Diversity (CBD), Dr Ahmed Djoghlaf. NParks, in collaboration with the global biodiversity community, formulated the Singapore Index on Cities' Biodiversity. The Plan of Action on Sub-national Governments, Cities and other Local Authorities for Biodiversity that lists the Singapore Index as a monitoring tool was endorsed by 193 parties at the 10th Meeting of the Conference of Parties to the CBD.

The Singapore Index comprises two sections, the city's profile and the scoring of the indicators. The indicators are grouped into three components: firstly, the native biodiversity in the city, secondly, the ecosystem services provided by the biodiversity in the city, and thirdly, the governance and management of the city's biodiversity.

Dr Lena Chan, director of the NBC, coordinated the international technical task force that guided the development of the Singapore Index. "The main purpose is not to make comparisons but for cities to monitor their own biodiversity efforts over time and to use the indicators as sources of inspiration to improve their biodiversity," she said. "As it is a form of self-assessment, we designed it to be easy to apply, but it has to be scientifically credible, objective and fair."

Currently, over 70 cities around the world are in various stages of applying the index.

Species named after Singapore

Daisy Sponge
(*Coelocarteria singaporensis*)

Singapore Vinegar Crab
(*Episesarma singaporense*)

Syndyas singapurensis

Bulbophyllum singaporeanum

Singapore Kopsia (*Kopsia singapurensis*)

Plantain Squirrel
(*Callosciurus notatus singapurensis*)

Singapore Durian (*Durio singaporensis*)

Species that were first discovered in Singapore but also occur elsewhere

Banded Leaf Monkey (*Presbytis femoralis*)

Estuarine Seahorse
(*Hippocampus kuda*)

Mangrove Pit Viper
(*Cryptelytrops purpureomaculatus*)

7

CITY IN A GARDEN

More than a "garden city", where the main function of greenery is to be decorative, Singapore is becoming a "city in a garden", an all-embracing living and working environment. Where space is short, green has spread to walls and rooftops. The revolution is "blue" as well as "green", involving rivers and canals. A network of connectors and corridors has spread across the island, linking green spaces to create an alternative dimension, safe from the roar of traffic. And the public have joined the conversation, as ways are sought to reconcile development priorities with the need to conserve.

● Dusk falls over central Singapore: a view over Kallang Riverside Park.

Gardening in the sky

When space for gardens at ground level is limited, you go upwards. Showing the creativity that has increasingly characterised Singapore's greening efforts in the 21st century, planners and developers began to apply greenery to roofs.

Some of the benefits are practical. Greenery helps reduce the heat island effect, whereby built-up areas can be up to 4 degrees Celsius hotter than forests. It is a heat insulator, reducing air-conditioning costs. It filters the air. It acts as a "sponge" for rainfall, reducing the danger of flooding. Properly installed, it can reduce maintenance costs.

For HDB, rooftop planting began with multistorey carparks. Because older buildings were not designed to take the weight of conventional gardens, HDB came up with a patented lightweight tray system, which uses little substrate and requires low maintenance. Also known as the Prefabricated Extensive Green (PEG) Roof system, with the right selection of plants it can thrive without irrigation or rain for up to three weeks.

Architects designed more ambitious rooftop gardens into new housing developments, as at Sengkang and Punggol, that are also recreational spaces for residents. For sustainability, the designers choose plants that are able to tolerate exposed dry conditions, are not too heavy and are easy to maintain. Trees cannot be too tall and must be suitable for growing in planter boxes.

URA introduced guidelines encouraging private developers and residents to provide balcony planting, communal planter boxes, landscaped decks, sky terraces and roof gardens. Skyrise greenery has proven to be popular among developers and building owners, helping to replace ground level greenery in land-scarce Singapore.

● Top: Rooftop greenery at Liang Seah Place, a restored row of heritage buildings.

● Above: The rooftop garden at New Town Primary School.

"Physically, NParks will have to...look for innovative ways to provide for and integrate greenery into a more densely built environment. Developers can and should incorporate more vertical and rooftop greenery into their buildings to soften the landscape."

Khaw Boon Wan, minister for National Development

A wall of green

Plants on vertical walls improve the look of modern buildings and help to protect them from the weather. They provide sound insulation. They provide a habitat for birds, butterflies and other small animals. Above all, they have a psychological benefit, helping to make the urban environment less stressful. Vertical vegetation is usually planted in modular cassettes or planters attached to a trellis or framework. For example, cassette planting is being used on some of LTA's ventilation blocks for the new North-South and Marina Coastal expressways.

Top: Vertical planting at Singapore Management University.

Centre: The rooftop garden at Orchard Central shopping mall, offering breathtaking views of the city.

Above left: Green wall overlooking the pool at Newton Suites condominium.

Above right: Vertical planting at Orchard Residence.

The island-wide water garden

Singapore has 32 major rivers and more than 8,000 km of canals and drains. Two-thirds of the island is now, in effect, water catchment. This was an opportunity to turn this functional infrastructure into a star feature of the garden that Singapore was becoming. The Active, Beautiful, Clean Waters (ABC Waters) programme, introduced by national water agency PUB, transforms this pervasive network of water bodies into beautiful and clean streams, rivers and lakes for the community to enjoy. Over 20 projects were implemented by 2012. More than 100 potential locations have been identified, while another 32 projects have been implemented by other public agencies and private developments.

The programme was done jointly with the development of the Park Connector Network and new parks and housing estates, where water was integrated from the beginning. It created more community spaces for people to enjoy recreational activities. The programme also had a positive impact on Singapore's biodiversity – some rare and uncommon birds have been spotted within urban parks and waterside habitats.

Across Singapore, PUB is also promoting the development of rain gardens, bioretention swales and wetlands, which are not only attractive but also natural filters, slowing down rainwater runoff and improving its quality.

"Our objective was to bring people closer to water, to build a relationship, so that they would think about water wisely and look after the catchment. We looked at our canals and drains and figured out how to beautify them, bring back nature and enable people to enjoy them."
Chew Men Leong, chief executive, PUB, the national water agency

Lorong Halus Wetland

A portion of the old Lorong Halus landfill along
the eastern bank of Serangoon Reservoir has
been transformed into an educational site and a
sanctuary for plants, birds and other wildlife. Using
a bio-remediation system, PUB designed the Lorong
Halus Wetland to collect and treat water passing
through the former landfill, preventing it from flowing
into Serangoon Reservoir. At the same time, the
wetland is a new haven for biodiversity and provides
opportunities for recreation, research and education.

Reservoir in the city

In 2008, the Marina Channel was dammed up to form the
Marina Barrage, creating a freshwater lake known as the
Marina Reservoir for Singapore's water supply. The barrage
also helps alleviate flooding in low-lying parts of the city
centre and provides an ideal venue for recreational activities.

This downtown reservoir is the centrepiece of attractions
around Marina Bay, hosting sports like dragon-boating and
kayaking. It also serves as a spectacular backdrop for major
events such as the Formula One Grand Prix and festive
celebrations.

> "In terms of how we want Singapore to be remembered by Singaporeans as well as by visitors, I think that Gardens by the Bay adds something very special to our metropolis."
> Lee Hsien Loong, prime minister

Gardens by the Bay

June 2012 saw the opening of Bay South, part of Gardens by the Bay, a vast "people's garden" designed to educate and entertain. Bay East and Bay Central (yet to be developed) complete the picture. Here, the public can see a diversity of plants from all over the world.

The decision to spend the money required to develop Gardens by the Bay was courageous. Singapore Botanic Gardens was suffering visitor overload, a victim of its own success, and there was room for another destination for plant lovers. But the most important motive was to communicate the magic of plants to the public. As Kenneth Er, chief operating officer, Gardens by the Bay, explains, those behind the project had Singaporeans in mind. "If you want to run any attraction successfully, you make sure that, first and foremost, the content attracts people from the local catchment area. No doubt tourists will come and that's welcome, but it's a bonus."

A secondary factor was the wish to reinforce Singapore's appeal internationally, now that Marina Bay was being developed and attracting international companies.

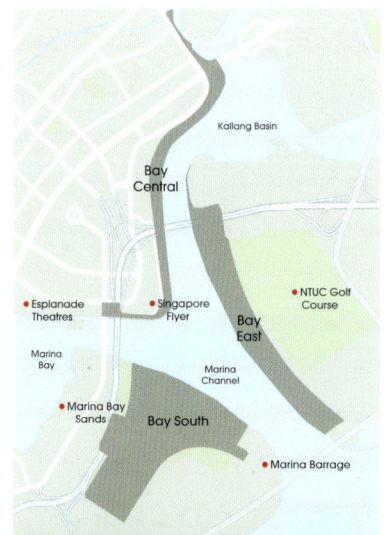

● Above: Spanning 54 hectares, Bay South is defined by its iconic attractions, the Flower Dome, Cloud Forest and Supertrees.

● Opposite, below left: The themed gardens, such as the Chinese Garden seen here at sunset, bring to life the importance of plants to people and the planet.

● Opposite, below right: Gardens by the Bay will occupy 101 hectares of prime land by the water; a gem on the "necklace of attractions" at Marina Bay.

Sustainability through technology

Gardens by the Bay was designed with sustainability in mind. The cooling technology in the glasshouses reduces energy consumption by at least 30 percent, as compared to conventional systems. The conservatories, such as the Flower Dome (below left), use special glass to reduce heat intake; the roofs automatically dispense "sails" to provide shade when necessary. The lower spaces of the conservatories are cooled by chilled water pipes embedded in the floor slabs. Electricity is generated on-site by a steam turbine fed by burning horticultural waste brought in from Singapore's parks and gardens. As the gardens lie next to a public reservoir, rainwater is channelled into the lakes; plants within act as a natural filter to clean it. Water saving is a priority; for example, mist is released in the Cloud Forest (below right) only where people and plants are.

● Above: The OCBC Skyway shows off the Supertrees with vertical gardens and offers a panoramic view of the Gardens and Marina Bay. Solar panels atop the Supertrees power their evening lighting.

"It is difficult to imagine Gardens by the Bay anywhere else – if you had winter and other seasons it would not be sustainable. We can sustain it throughout the year, with a continuous display of flowering by plants from north and south of the equator. We have perpetual summer."

Dr Kiat Tan, chief executive officer, Gardens by the Bay

● Top: Bay South is surrounded by the Dragonfly and Kingfisher Lakes. Water is collected in the lakes and filtered by aquatic reedbeds and plants.

● Below left: Students gather before the "Planet" sculpture by British artist Mark Quinn. Sculptures around the Gardens complement the beauty of the plant displays.

● Above: Rhododendrons (left) and fuchsias (centre) bloom in the climate-controlled cooled conservatories. At a pond, a Yellow Bittern (*Ixobrychus sinensis*) stalks prey (right).

● Below right: A couple pose for wedding photographs along the Dragonfly Lake boardwalk. Bay South's attractions are balanced with many quiet spaces for reflection.

Sustaining the Singapore garden

Singapore was ranked 52nd out of 132 countries in the 2012 Environmental Performance Index, placed in the "modest performers" group. This index is produced by Yale and Columbia Universities in collaboration with the World Economic Forum and the European Commission.

This placing is better than it might seem at first glance. The index compares Singapore, a city-state, with mostly larger countries with rural hinterlands. There is not much potential in Singapore for wind energy, for example.

However, as long ago as 1992, the government produced the Singapore Green Plan, taking into account input from a wide range of interest groups and stakeholders, including NSS. It was updated 10 years later by the Singapore Green Plan 2012 and revised again in 2006. That itself was superseded in 2009 by the Singapore National Biodiversity Strategy and Action Plan, coordinated by NParks, and "A Lively and Liveable Singapore: Strategies for Sustainable Growth", published jointly by the Ministry of the Environment and Water Resources and Ministry of National Development.

Progress has been made. For example, air quality, as measured by the Pollutant Standard Index, has been in the "good" range for at least 85 percent of the period since 2003, largely due to control of industrial and vehicle emissions. The government encourages Singaporeans to use public transport more. The MRT network is planned to double over the years 2008–2020. The bus fleet is set to grow. As Chew Hock Yong, chief executive of LTA, said, "People set high standards for us and it spurs us to do better."

Through funding and incentive schemes, the government promotes energy efficiency, clean energy, green buildings, water and environmental technologies, green transport and shipping, and waste minimisation. As Deputy Prime Minister Teo Chee Hean, chairman of the Interministerial Committee on Climate Change, said in 2012, "Everyone has a part to play."

Living in the future

As Singapore continued to urbanise, land constraints and increased awareness of green issues triggered a period of innovation in which many agencies played a role. It was a matter of putting Singapore's resources to creative use, with an emphasis on sustainability.

For example, HDB decided to develop Punggol as an Eco-Town: "The Sustainable Waterfront Town in the Tropics". Punggol Eco-Town is positioned as a "living laboratory", to test-bed and develop new knowledge in the areas of urban planning, environmentally-friendly design and green building solutions for a sustainable living environment. These solutions would be scalable and replicable in other estates and towns if proven viable.

At the same time, due to the need to increase the water catchment areas in Singapore to cater to increasing water needs, engineers at PUB decided to dam two rivers, Sungei Punggol and Sungei Serangoon, and form two reservoirs. A pipeline link between them was planned, to allow the transfer of water between the two reservoirs and to control their water levels. It was at this juncture that the opportunity for the creation of a new and unique housing environment in Singapore was seized, whereby water could be brought closer to the residents and a new waterfront living experience could be created. Led by HDB, with the collaboration of LTA, NParks, PUB, SSC (Singapore Sports Council) and URA, the idea of a landscaped waterway weaving through the town was conceptualised and implemented under the "Remaking Our Heartland" programme in 2007.

● Above: The Sunrise Bridge at Punggol Waterway, where visitors can take in the scenic views of the sunrise and surroundings.

● Left: Enjoying the coolness of the evening at Punggol Point Walk.

Treelodge@Punggol

Well-shaded walkways link the housing blocks that make up Treelodge@Punggol. The buildings have been oriented and spaced to minimise heat gain and optimise wind flow, not only at the precinct as a whole but also at the level of building blocks and individual flats. Some flats come with balconies or planter boxes. The roof of the carpark is designed as an Eco-Deck with lush greenery, leisure amenities and a community garden. Building columns are clad with vertical greenery. Rainwater is harvested for washing of corridors and for watering plants. The centralised chutes for recyclables, fitted to every block, have led to a threefold increase in recycling.

No winners, no losers

In earlier times, some older park managers recall, there was a tendency on the part of many policy-makers to see conservation and economic development as incompatible – you had to choose one or the other. Although development decisions can still be difficult, from the 1990s onwards, the debate became more thoughtful. For Prof Leo Tan, chairman of the Garden City Fund, the new mindset was encouraging. "I'm most proud of Singapore's self-discovery. There is a move to preserve nature in order for us to have a sustainable future." Ng Lang was CEO of NParks before he became URA's CEO and has seen the world from both points of view. He believes that Singapore benefited from this more consensual approach. "Greening has been deeply embedded into planning. It is part of a proactive process to make the living environment better, rather than an aspect of environmental activism."

The tranquillity of Raffles Terrace, Fort Canning Park.

"With strong capabilities properly institutionalised, NParks is ready to build on its strong foundation to take our Garden City to its next level of excellence. As Singapore will continue to urbanise, further urbanisation must not be at the expense of our quality of life."

Khaw Boon Wan, minister for National Development

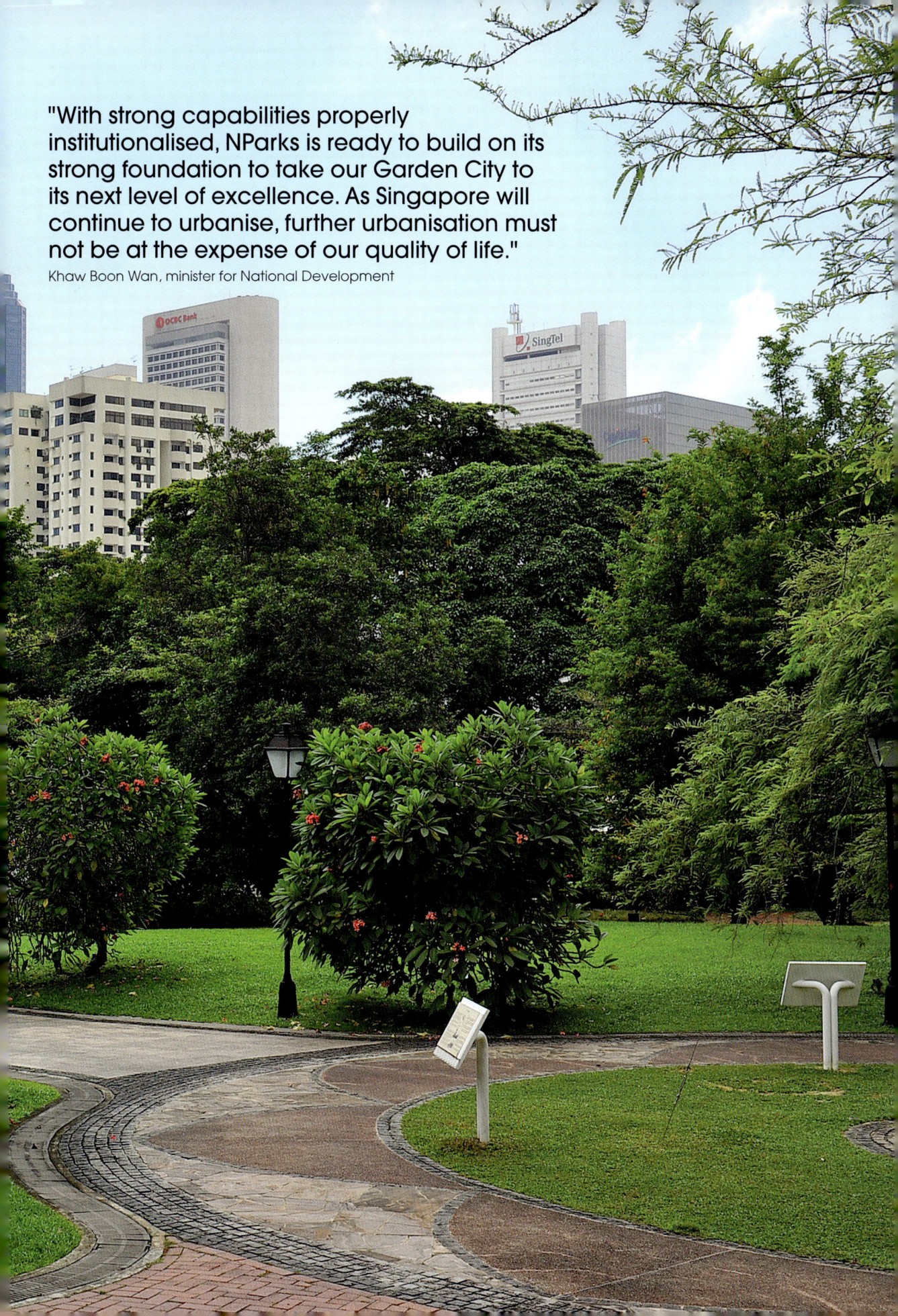

We need a plan

Land use planning has been essential to Singapore's development since the time the government drew up the first master plan in 1958. A key moment in the modern planning story came in 1971, with the completion of the first concept plan – a strategic land use and transportation plan mapping Singapore's development 40-50 years into the future. This plan, developed with United Nations assistance, included the development of satellite towns and allocation of land for parks. URA updated the concept plan in 1991 and 2001. The 1991 Concept Plan focused not only on economic growth but also on quality of life – it proposed greater variety and choice in housing and recreational facilities. Through extensive public consultation and coordination between agencies, the 2001 Concept Plan mapped out a vision for Singapore as a thriving world-class city. One key focus of the 2001 plan was on increasing the amount of green space for residents to enjoy. The successive plans reflected the increasing importance attached to greenery and conservation as time went by. In 2002, two additional nature reserves, Labrador Nature Reserve and Sungei Buloh Wetland Reserve, were gazetted.

URA went further, to recognise and safeguard areas with rich biodiversity through the inclusion of a Parks and Waterbodies Plan in 2002. The plan proposed more areas of natural beauty, waterfronts and beaches, developing the Park Connector Network, creating rooftop, vertical and balcony gardens, and furthering the greening of streetscapes. All these plans were the outcome of intense collaboration between government agencies and the public. As part of Master Plan 2008, the Parks and Waterbodies Plan was updated, and an island-wide Leisure Plan unveiled proposals for the Round Island Route and new nature destinations in the Southern Waterfront.

● Nature enthusiasts on a trail in the Central Catchment Nature Reserve.

"The challenges we face today are a little different than before. Given the pressures of our small geographical size, growing economy and population, GCAC is looking for new and innovative ways of greening to mitigate the effects of urbanisation and densification so that Singapore will continue to thrive as a highly liveable city."

Chang Hwee Nee, deputy secretary (planning), Ministry of National Development, and chairman, Garden City Action Committee

The little green dot

Satellite imagery shows that the amount of Singapore's land area covered by greenery was over 40 percent in 2011. The colours in this map indicate degrees of vegetation – light and dark greens indicate the density of vegetation. Some green cover is on state land, which may, if circumstances demand, have to be used to meet development needs.

How NGOs can play their part

Modern society gives an important role to NGOs such as the Nature Society (Singapore), whose members are passionate about nature and conservation. As use of the Internet became universal, many other formal and informal organisations were formed, encouraging people of all ages to pursue their interest in nature, and through blogs or other social media express their opinions.

NParks was a facilitator, encouraging bodies such as the NSS to play a more positive role in making the case for protecting Sungei Buloh, Labrador and Chek Jawa. This way, the government became aware of the public's attachment to green spaces and conservation, and the public gained some ownership over the way space was managed. Conservation issues became headline news, matters of public interest. In this new atmosphere, NSS was part of the mainstream, sharing its knowledge with the government, universities and commercial companies, so that planning decisions could be made in the best interests of Singapore.

Ng Lang, CEO of URA, sees the relationship between his organisation and nature groups as a partnership. "It has helped shape the garden city," he said. "It is more productive than the confrontational situation in many other cities, where activists advocate and officials stonewall." Former president of NSS Dr Geh Min agrees: "Nothing is more educational than encouraging members of the public to think over problems of land scarcity with the government, rather than in a conflict situation."

"(the NSS) is a valuable pressure group in society. They push for their interest and we take their views seriously. We cannot always accommodate everything they would like us to do in an ideal world, but they remind us what it is that we are giving up when we have to make a choice. And I think we have a good working relationship with many of these groups. We would be the poorer without such passionate people."

Lee Hsien Loong, prime minister

"It is imperative that all parties – the public agencies, academics, NGOs and corporations – work together. Where our NParks colleagues shine is that they are adept at facilitating and encouraging collaboration."

Dr Leong Chee Chiew, commissioner of parks and recreation, NParks

The Internet age

In the digital era, the leadership of the green movement is "fuzzy", according to Prof Peter Ng, director, Raffles Museum of Biodiversity Research. "Groups pop up when they feel like it. People become very committed to particular areas and gather around a natural leader to focus on, say, coral reefs or rainforests. Then they blog and tweet – social media have become pervasive. That being said, the social media, the digital age in general, have been very good for environmentalism."

Information is easily available. Established organisations are no longer the only, or even the primary, source. Not only are there good books on flora and fauna, but anyone can go to the computer, type a few words into a search engine and get a good photo of almost any interesting plant or animal. Data gathered on digital cameras and handphones spread incredibly fast.

NSS president Dr Shawn Lum sounds a cautionary note. "The challenge is to get people to see for themselves. Being a virtual naturalist, looking at everything on the computer, is one thing. When you ask people to get out and see things literally on the ground, many of them can get a bit hesitant."

Enthusiasts can set up school nature societies and Facebook groups – these bring people together to do things they could not do alone. Ecology researcher Marcus Chua suggested several project ideas: "They could do surveys of flowering of plants and wayside trees in Singapore...; what birds can be found in each housing estate at what times of the year. Anyone can set up a database and submit information via cloud sharing. Animal records can be collected – it would be really good to get people collecting records for flowering seasons or seasonal

● Above: The internet and gadgets such as tablets and smartphones play a central part in the lives of Singaporeans, young and old.

● Below: Children take a photo of an earthworn using a smartphone.

bird visits. On a large scale it would be wonderful."

Although NSS takes part in public debate on conservation issues, Dr Lum, believes that simple love of nature is still what motivates its members to take part in nature walks, workshops, short courses and eco-trips. "If more of our wild areas were secure, our members would still be mostly bird-watching and keeping records of butterflies and plants. We are still hobbyists at heart."

Engaging the community online

In addition to the NParks website, people interested in knowing more about the organisation or the work it does can visit a few sites. Flora and Fauna Web (below) (florafaunaweb.nparks.gov.sg) is a one stop information portal for the knowledge-hungry nature lover or hobby gardener. It provides detailed information on a huge range of terrestrial, freshwater and marine animals to be found in Singapore, as well as plants, both native and exotic, grown in the region.

Then there is the NParks Facebook page (bottom) (www.facebook.com/nparksbuzz), where people can share photos or stories. Occasionally, photos of wildlife or flowers are posted with a request for identification.

Twitter (www.twitter.com/nparksbuzz) is also a way NParks shares information with the connected community.

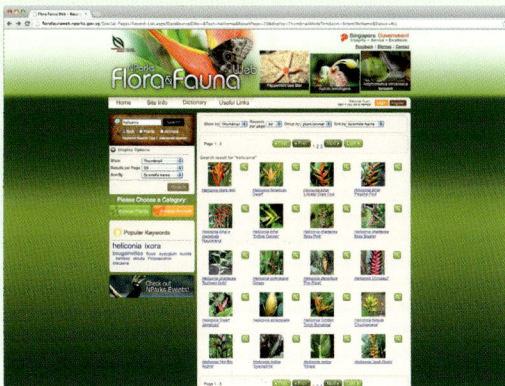

NParks – Let's Make Singapore Our Garden updated their cover photo.
April 1

Thank you for sharing this photo with us, Mike To. We were moved by this!

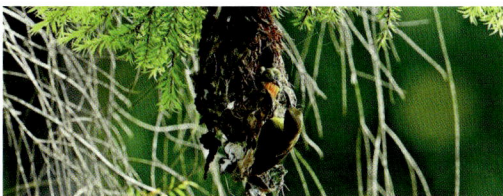

Like · Comment · Share

87 people like this.

Mike To You are most welcome, let's work together Make Singapore Our Home & Garden.
Like · Reply · 2 · April 1 at 9:36am

Hamzah Osman We have this in Teck Ghee Primary school too after i grew so many Butterfly Pea along school fence.
Like · Reply · 1 · April 1 at 9:46am via mobile

Ian Goh I have seen this in my and my neighbour's backyard. Sad thing is with newer URA permission to allow landed houses to build nearer to its boundary at the expense of gardens, I fear soon we will no longer see them again.
Like · Reply · 1 · April 1 at 10:03am

Patrick Wang Olive back sunbird
Like · Reply · 1 · April 1 at 10:09am via mobile

Websites to visit

There is a huge range of local and global websites devoted to nature and ways in which anyone can enjoy it, learn about it and play a part in conserving it. To take just two examples, wildsingapore.com (below) is a collection of web resources brought together by an individual, Ria Tan. It started out as a companion website to a Chek Jawa guidebook. Since then, it has grown to contain a vast amount of information on "wild places, wild activities and wild people".

Led by Khew Sin Khoon, a group of butterfly enthusiasts formed ButterflyCircle (bottom), with a focus on butterfly research and photography. The group then set up a website, forum and blog to facilitate discussions, as well as to record and share information on butterflies in Singapore.

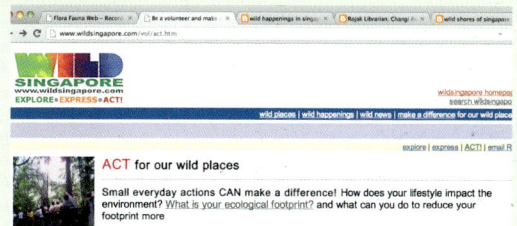

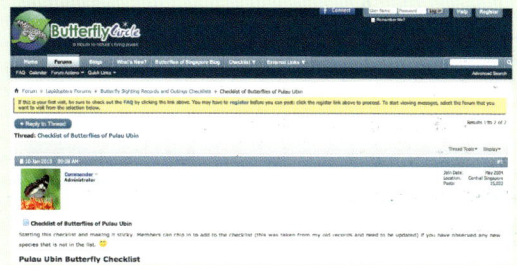

sParks*

Moving with the times, through a Public Private Partnership programme, NParks has introduced sParks*, a map-based app for mobile devices, with directions to parks, park connectors and nature reserves nearby. It features events and activities, and makes it possible to create networks of friends with shared interests.

Reaching out, enriching lives

When NParks was formed, one of its goals was to get people interested in plants and animals, so that they would appreciate parks and greenery. One way was to get nature into the school curriculum and to produce educational materials for students of all ages. NParks introduced "programming" to parks – this is the professional term for putting on a programme of exhibitions, shows, visits and other events. They formed children's groups and created appealing characters to bring the subject alive. No one who saw her could forget Sarah the Botanicosaurus, who lost her squirrel friends in the Evolution Garden and called on children to help her find them again.

The range of activities grew and grew, to include programmes for "young naturalists", themed walks, practical workshops and training for teachers – not to mention events for nature-loving adults too.

● Top: A school group at the Pasir Ris Mangrove Boardwalk.

● Left: "Lost in the Garden City" is a play conceptualised by NParks that educates schoolchildren about greenery and biodiversity.

"Nature Cares, which brings corporate sponsors, school groups and community groups together through nature-based activities, goes one step further. Not only do the students bring cheer to the lives of the elderly, but they also get to take part in nature-related activities, which have physical, mental and social benefits."

Grace Chua, principal, Queenstown Primary School

● Above: Children participating in NParks' Kids for Nature programme, which is part of the Ministry of Education's Programme for Active Learning.

● Below: Nature Cares programme volunteers helping members of the Lee Kong Chian Gardens School (MINDS) (left) and residents of St Theresa's Home (centre).

● Below right: As part of the Every Child a Seed initiative, more than 41,000 students in the Primary 3 cohort received a planting kit.

Getting involved

Making a difference

The Garden City Fund (www.gardencityfund.org) was set up in 2002, so that individual donors and companies could play a more active role in caring for our natural heritage. To date, it has enabled greater community outreach to individuals and corporations. Among the projects made possible by such partnerships are the Ubin-HSBC Volunteer Hub, SPH Walk of Giants, outdoor classrooms and surveys.

Ever since 1993, NParks has encouraged people of all ages to join volunteer programmes. It started that year with an informal but very keen group of bird-watchers at Sungei Buloh. Three years later, an organised training scheme was set up for volunteers who wanted to take part in conservation activities there. Later, similar programmes started at other sites, such as Singapore Botanic Gardens, the Central Catchment area and Pulau Ubin. Today, NParks has more than 800 volunteers, from nine years old to over 80, who help visitors, act as guides, join in community gardening, and participate in conservation and marine biodiversity surveys. Recently, NParks showed schoolchildren how to recognise and remove alien plants that had invaded the rainforest at the Singapore Botanic Gardens.

Volunteers can help society too. Studies have shown that experiencing green environments can improve mental health and well-being. NParks launched the Nature Cares programme in 2012 to encourage corporations and children to serve the needs of the community through nature-based activities. Volunteers would spend time with community groups, including the elderly and children with special needs. Another programme, Community in Nature, was an initiative for the engagement of local communities through volunteerism and collaboration with NGOs.

"A garden can be a habitat for wildlife as well as a collection of plants if it is actively managed. It can function as an ecosystem and serve an important conservation purpose while still being man-made."

Dr Shawn Lum, president, Nature Society (Singapore)

● Opposite: Nature artist and volunteer Tham Pui San has been conducting free art workshops at Sungei Buloh Wetland Reserve every month for several years.

● Above: Comprehensive Marine Biodiversity Survey volunteers conducting surveys at the mudflats of Kranji Nature Trail.

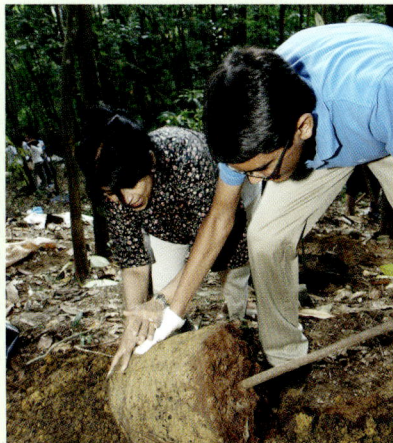

Plant a tree!

The aim of the Plant-A Tree programme is to give everyone the chance to make a contribution to our greenery and plant a tree personally. The sapling can be planted during one of the monthly tree-planting days organised by NParks and the Garden City Fund. In early 2013, NParks announced an initiative to plant 1,963 trees between June and November, to mark that day when Lee Kuan Yew planted a tree 50 years earlier.

"Sensitively planned, intense urbanisation and a high-quality living environment can be synonymous."

Khaw Boon Wan, minister for National Development

Now, 50 years on ...

The lush and beautifully maintained greenery of Singapore lives up to its international hype. There was a time when it was "added on" and was perhaps a bit superficial. Now it has become a well-known national characteristic of Singapore, a place so built up and yet so natural. Singapore's geographical position endows it with a warm and wet climate with virtually no seasons. There is no better place to grow a vast range of plants introduced from around the world, to complement the riches of the native biodiversity. And as a result, there is no better place to live, many would say.

This has been the story of a 50-year transformation. Singapore would not have become a city in a garden without the efforts of individuals, corporations and agencies all pulling in the same direction, convinced that making Singapore an appealing place to live and work was important. People working on planning, transport, housing, industrial development, water resources, education, scientific research – all have played a part. Driving the process are the men and women creating and looking after the parks and gardens and streetscape greenery, whether as managers, arborists, horticulturists, botanists, entomologists, ecologists, landscape designers, cleaners or workers pruning trees. However, the progress of the country's parks, gardens and nature reserves over the next 50 years will depend on the support of the public.

"Moving forward, we hope that more Singaporeans can step forward, roll up their sleeves and join us in creating our City in a Garden."

Poon Hong Yuen, chief executive officer, NParks

Acknowledgements

This book has been made possible with the generous sponsorship of

Gold Sponsors

Dairy Farm Singapore

ENGRO BUILDING SUSTAINABILITY

Far East Organization INSPIRING BETTER LIVES

TEMASEK

Silver Sponsors

TUAS POWER A member of China Huaneng Group

WOH HUP (PRIVATE) LIMITED BUILDING WITH INTEGRITY

The NParks editorial team would like to thank all NParks colleagues both past and present, partners, volunteers, and supporters who have assisted in this project, from taking the time to be interviewed, to digging up photos from their archives, to verifying information and for helping in a myriad other ways.

Photo Credits

Images not listed below are provided by the National Parks Board, the copyright holders. All other photographs are reproduced courtesy of the following and may not be used without the written permission of the copyright holders.

Ahmad Iskandar bin Abdullah: 178–179
Alan Yong: 18–19
Alisa Wee: 96–97
Ang Chwee Leng: 94 (bottom)
Ang Wee Foong: 133 (bottom), 167 (bottom left)
Antonius Mak Wai Kwong: 147 (top)
Benjamin Lee: 134, 135 (top left)
Billycane Lim Gim Hooi: 161 (inset (second from bottom))
Boo Ghim Yew: 75 (bottom)
Brice Li: 156–157
Cai Yixiong: 159 (middle right), 166 (top left and bottom right)
Chan Jin Kiat Arik: 80 (top left)
Changi Airport Group: 43 (top left and top right)
Prof Cheah Jin Seng: 22 (top)
Cheah Kin Wai: 28–29
Chen Lilian: 158
Cheong Jia Jun: 159 (middle row, left)
Cheryl Chia: 167 (bottom centre)
Chew Bee See: 70–71 (gatefold)
Chia She Jian: 100 (bottom)
Chin Kwee Soon Patrick: 159 (bottom left)
Ching Whee Main: 6–7
Chong Kee Wah: 124–125
Chow Kian Yew: 95 (centre)
Christine Luo Wenyi: 136 (top right)
Chua Ee Kiam: 160–161
Clarence Chua: 39 (top left), 80 (bottom left)
Danny Teo: 152 (left)
Darric Tan: 105 (top)
David Ng Soon Thong: 145 (bottom right)
Dylan Lim Yi Wei: 181 (bottom right)
Edgar Lim: 145 (bottom (inset))
Elmich Pte Ltd: 171 (centre)
Felicia Wei: 78 (top)
Gardens by the Bay: 174 (bottom right), 175 (bottom right), 177 (bottom left)
Genevieve Ow: 81 (top and bottom left)
Geoffrey Davison: 155 (bottom right)
Goh Gan Khing: 36 (top left), 37 (top left)
Gretchen Liu: 22 (bottom left)
Hadi Azhari Ishak: 2–3
Han Cheng Yuen: 53 (middle row, left)
Han Xuan Yan: 159 (top left)
Housing & Development Board: 45 (top)
Huen Kwong Leung David: 167 (top centre)
Jamsari Ahmad: 184–185
Jeremy Ang: 127 (bottom right)
Jervis Mun: 175 (bottom left)
Jocelyne Sze Shimin: 130 (top)
Joey Yu Ziyi: 167 (middle row, right)
Joseph Goh Meng Huat: 69 (top)
JTC Corporation: 46, 47
Julia Anderson: 138 (bottom centre and bottom right), 139 (centre)
Karenne Tun: 162 (inset)
Kenneth Er: 174–175 (top), 174 (bottom left), 176, 177 (top, middle row and bottom right)
Kevin Fan Kei Yim: 159 (top right)
Khew Sin Khoon: 166 (top right and bottom centre)
Kirk Benjamin: 12–13
Koh Poo Kiong: 91 (bottom)
Land Transport Authority: 33 (bottom)

Lee Jia Hwa: 40 (top)
Lena Han: 193 (top)
Leong Kwok Peng: 187 (bottom)
Leow Yue Chong Francis: 171 (bottom right)
Lim Chin Pong: 146 (top right)
Lim Fung Yenn: 130 (bottom right)
Lim Shen Jean: 39 (bottom)
Lionel Gouw Kian Hui: 133 (centre), 146 (left, second from top)
Lok Yan Ling: 170
Loke Peng Fai: 159 (bottom right)
Low Choon How: 4–5
Marcus Chua: 139 (top left)
Marina Chekunova: 133 (top right)
Masita Bakti: 95 (top)
Mendis Tan: 144 (bottom (inset)), 145 (top), 147 (centre), 167 (bottom right)
Ministry of Communications & Information Collection, courtesy of National Archives of Singapore: 88 (top)
Mohamad Azlin bin Sani: 188 (top)
Mohd Imran bin Mohd Kefli: 130 (inset)
Myron Tay: 135 (top right)
Nanthini Elamgovan: 150 (bottom)
National Heritage Board (gift of Mr G.K. Goh): 22 (bottom right), 23 (top)
Neo Mei Lin: 162 (background), 165 (middle row)
Ng Jia Wei: 118 (top left)
Ng Mui Choon: 146 (bottom)
Ng Siew Boon: 8
Noel Thomas: 78 (bottom), 138 (top), 139 (top (second from top)), top far right), 161 (inset (second from top))
Oei Geok Baw: 107 (bottom)
Ong Guat Pheng Shirline: 102
Ong Kian Chai: 89 (top)
Ong Kim Liap: 43 (bottom right)
Ong Zhen Quan: 133 (top left)
Patrick Grootaret: 167 (top right)
Phua Kia Wang: 154 (left)
Dr Pim Sanderson: 31 (bottom right)
PUB, The National Water Agency: 98 (bottom and centre), 172–173
Raffles Hotel Collection: 54 (top)
Rene Ong: 163 (top right and middle row, left)
Ria Tan: 127 (bottom left), 136 (top left), 140 (bottom), 163 (top left), 167 (top left)
Richard Seah: 110 (top)
Robert Teo: 137 (bottom), 187 (top), 193 (bottom left)
Rushinah binte Idris: 128 (top)
Samson Tan: 86–87
Seet Hui Ying: 153 (centre)
Serena Lee: 69 (bottom)
Shannon Heng: 137 (top left)
Shee Zhi Qiang: 27 (bottom right), 44
Shereen Tan: 188 (bottom)
Shirley Ng: 167 (left (second from bottom))
Shirly Hamra Hong: 113 (top)
Sia Fook Kee Johnson: 137 (top right), 159 (middle row, centre)
Singapore Botanic Gardens archives: 22 (bottom centre), 23 (bottom)
Singapore Press Holdings: 16, 24, 25 (top), 26, 27 (top right), 61 (top), 103 (middle row, left)

Sng Yue Jin: 149 (top)
Tai Li Jeng: 147 (bottom)
Takako Imai: 155 (bottom left)
Tan Chee Hiang: 111
Tan Heok Hui: 166 (bottom left)
Tan Tuan Khoon: 53 (top left)
Tan Tze Siong: 161 (top (inset))
Tee Swee Ping: 30, 31 (top right and bottom left), 32 (top right and bottom left), 34 (right), 35, 36 (bottom left), 37 (top right, bottom), 143
Teng Jee Hiem: 168–169
Teo Siyang: 32 (bottom right)
Tey Boon Sin: 154 (right)
Tham Pui San: 192
Timothy Auger: 42, 74 (bottom), 92 (top), 142 (top left), 163 (bottom)
Timothy Lee: 94 (top right)
Tor Peng Hock: 141
UK National Archives: 21
Wong Chee Yen: 153 (top)
Wong Tuan Wah: 10–11, 20, 49, 50–51 (gatefold), 53 (bottom), 54 (bottom), 55 (top and bottom), 126, 131 (bottom left), 132, 144 (top), 199
Wong Yan Zhi: 155 (top)
Wong Yew Kwan: 116 (bottom)
Dr Yam Tim Wing: 41 (bottom), 167 (second from top left)
Yeap Khek Teong: 120–121
Yong Gek Chin: 106

The satellite map on page 185 was produced by DHI Water & Environment Pte Ltd by processing satellite data acquired and processed by Centre for Remote Imaging, Sensing and Processing (CRISP), National University of Singapore. The method and algorithms used in producing the map were developed by CRISP.

Every effort has been made to trace the copyright holders and we apologise for unintentional omissions. If any due acknowledgment of copyright clearance has been inadvertently overlooked, the publisher will be pleased to make good the omission in subsequent printing.

The information in this publication was accurate at the time of printing.

References

Description	Data	Reference
Vascular plant species	3,971	Chong, K.Y., Tan, H.T.W. & Corlett, R. (2009). *A checklist of the total vascular plant flora of Singapore: Native, naturalised and cultivated species.* Singapore: Raffles Museum of Biodiversity Research and Department of Biological Science. p. 273
Of 3,971 vascular plant species: • Native vascular plant species • Exotic vascular plant species	2,145 1,826	Chong, K.Y., Tan, H.T.W. & Corlett, R. (2009). *A checklist of the total vascular plant flora of Singapore: Native, naturalised and cultivated species.* Singapore: Raffles Museum of Biodiversity Research and Department of Biological Science. p. 273.
Lichen species	296	Sipman, H.J.M. (2009). Tropical urban lichens: Observations from Singapore. *Blumea.* 54: 297–299.
Mammal species	52	Baker, N. & Lim, K. (2008). *Wild animals of Singapore: A photographic guide to mammals, reptiles, amphibians and freshwater fishes.* Singapore: Draco Publishing and Distribution Pte Ltd & Nature Society (Singapore).
Bird species	364	Wang, L.K. & Hails, C.J. (2007). An annotated checklist of the birds of Singapore. *Raffles Bulletin of Zoology* Supplement 15. p. 179.
Reptile species	103	Baker, N. & Lim, K. (2008). *Wild animals of Singapore: A photographic guide to mammals, reptiles, amphibians and freshwater fishes.* Singapore: Draco Publishing and Distribution Pte Ltd & Nature Society (Singapore).
Amphibian species	28	Baker, N. & Lim, K. (2008). *Wild animals of Singapore: A photographic guide to mammals, reptiles, amphibians and freshwater fishes.* Singapore: Draco Publishing and Distribution Pte Ltd & Nature Society (Singapore).
Butterfly species	301	http://www.butterflycircle.org/sgchecklist.htm (as of 10 Jan 2013)
Dragonfly species	127	Robin Ngiam (pers. comm.)
Reef fish species	107	Low, J. & Chou, L.M. (1992). Distribution of coral reef fish in Singapore. In Chou, L.M. & C.R. Wilkinson (Eds.) (1992). Third ASEAN Science and Technology Week Conference Proceedings, Vol 6, *Marine Science: Living Coastal Resources,* 21–23 Sept 1992.
Freshwater fish species	66	Baker, N. & Lim, K. (2008). *Wild animals of Singapore: A photographic guide to mammals, reptiles, amphibians and freshwater fishes.* Singapore: Draco Publishing and Distribution Pte Ltd & Nature Society (Singapore).
Goby species	149	Larson, H.K., Jaafar, Z. & Lim, K.K.P. (2008). An annotated checklist of the gobioid fishes of Singapore. *Raffles Bulletin of Zoology.* 56(1): 135–155.
Hard coral species	255	Huang, D., Tun, K.P.P., Chou, L.M. & Todd, P. (2009). An inventory of zooxanthellate scleractinian corals in Singapore, including 33 new species records. *Raffles Bulletin of Zoology.* 22: 69–80.
Sponge species	About 200	Lim, S.C., Voogd, N.D. & Tan, K.S. (2008). *A guide to sponges of Singapore.* Singapore: Singapore Science Centre. p. 173.
Molluscs species	1,264	Tan, S.K. & Woo, H.P.M. (2010). *A preliminary checklist of the molluscs of Singapore.* Singapore: Raffles Museum of Biodiversity Research. p. 78.
Echinoderm species	90	Lane, D.J.W. & Spiegel, D. Van der. (2003). *A guide to sea stars and other echinoderms of Singapore.* Singapore: Singapore Science Centre. p. 187.
Seagrass species	12	http://www.seagrasswatch.org/Singapore.html Additional Information

⬤ Page 196–197: The Swan Lake was added to the Singapore Botanic Gardens as a feature in 1866. Overlooking the lake is a Victorian cast-iron garden shelter that once stood at Grange Road. In 1969, it was dismantled and re-erected in the Gardens at the entrance to the Rain Forest. It was moved to its current site in 2001.

⬤ Page 199: In 2013, the nest of a White-bellied Sea Eagle (*Haliaeetus leucogaster*) was spotted in a tree in Fort Canning Park. When the eaglets were born, they created quite a stir among NParks staff and park visitors.